Artificial Intelligence and the Rule of Law

Artificial intelligence is an important driving force of the new round of technological revolution and industrial change. Artificial intelligence and blockchain have risen to the level of national policy. While the rapid development of AI brings convenience to people's lives, it also poses challenges to global changes and the development of human society, social governance and legal regulation, such as the leakage of personal privacy, frequent data security problems, and prejudice and discrimination caused by algorithms. AI is profoundly changing the world, and it is necessary to think about how to promote the construction of responsible AI from the perspective of global governance in the future.

In response to these issues, this series aims to explore and study the new situations and new issues at the intersection of AI, law and political science all over the world, to provide a legal basis for the development of AI-related frontier fields and to limit the development of AI within the framework of law to promote the harmonious and orderly construction of AI and legal systems. Based on the above purpose, this series is selected and integrated with multidisciplinary strengths to strengthen the research on legal, ethical, and social issues related to AI, focusing on laws, regulations, institutional systems, and ethics for the healthy development of AI. The authors of this series are all top authoritative experts and scholars in AI, law, political science, and other disciplines in China, and they have already made considerable achievements and social reputations in this field of research.

This series has a clear awareness of the problem and proposes specific solutions for the implementation and implementation of AI governance from a multidisciplinary and multidisciplinary perspective, covering AI, blockchain, algorithms, digital currency and other specific segments closely related to law, with cutting-edge, academic and theoretical value, and filling the gap of similar research in China.

Qiqi Gao · Jiteng Zhang

Artificial Intelligence Governance and the Blockchain Revolution

上海人民出版社

Springer

Qiqi Gao
Institute of Political Science
East China University of Political Science
Shanghai, China

Jiteng Zhang
Institute of Political Science
East China University of Political Science
Shanghai, China

ISSN 2731-6238 ISSN 2731-6246 (electronic)
Artificial Intelligence and the Rule of Law
ISBN 978-981-99-9210-2 ISBN 978-981-99-9211-9 (eBook)
https://doi.org/10.1007/978-981-99-9211-9

Jointly published with Shanghai People's Publishing House
The print edition is not for sale in China (Mainland). Customers from China (Mainland) please order the print book from: Shanghai People's Publishing House.
ISBN of the Co-Publisher's edition: 978-7-208-16693-6

This Springer imprint is published by the registered company Springer Nature Singapore Pte Ltd.
The registered company address is: 152 Beach Road, #21-01/04 Gateway East, Singapore 189721, Singapore

Introduction: Artificial Intelligence, Blockchain, and National Governance Modernization

The understanding of *once-in-a-century changes* should be improved from the perspective of the four industrial revolutions. The ongoing Fourth Industrial Revolution can also be summarized as the intelligence revolution. In this revolution, the developing countries represented by China and the developed countries represented by the United States have formed a new era of competition with technology as the core. This situation can be seen as the technological dimension of the current global change. The Western developed countries led the three previous industrial revolutions. Britain led the First Industrial Revolution, Germany and the United States led the Second Industrial Revolution, and the United States led the Third Industrial Revolution. Previously, Western developed countries monopolized the dominance of the technological revolution. In contrast, in the Fourth Industrial Revolution, China competed with developed countries for the first time on behalf of developing countries. This is a significant background for Western developed countries led by the United States to worry about China's rapid progress in science and technology and even to contain China's technological development. Science and technology constitute a primary productive force. Western developed countries have been accustomed to the domination of science and technology in recent times, so they are unwilling to let developing countries gain the dominance of science and technology. Therefore, when developing countries show partial superiority in science and technology, Western countries will curb developing countries' scientific and technological progress through integrated methods, such as patent strategies and exchange restrictions. Together, these measures constitute the hegemonic activities of developed countries in science and technology. A series of U.S. hegemonic behaviors in the field of science and technology since the Sino-U.S. trade dispute is a clear testimony of this feature.

There are two critical technologies in today's smart revolution: intelligence technology and blockchain technology. Intelligence technologies include not only artificial intelligence technologies but also bio-intelligence technologies. Bio-intelligence technology, in turn, includes brain science, neurocognition, and life science-related technologies such as gene editing. In this sense, energy technology and material technology are also critical auxiliary technologies for the smart revolution. With the assistance of these technologies, the intellectualization of biological intelligence will be significantly improved, while the computing efficiency, resource consumption, and supporting hardware of artificial intelligence can be improved. The vast productivity generated by intelligent technology will provide the primary technical support to solve the problems of human poverty and development.

The second type of technology is blockchain technology. At present, the focus of blockchain technology is still mainly on information exchange between blocks, but in the future, communication between chains will become very important. The leading solution to achieve data exchange and interoperability between different blockchains is sidechain technology. In the long run, the "chain" of blockchain needs to form an interlocking and complex mesh structure that gradually constitutes the next-generation Internet, i.e., the value Internet.

Intelligence technology and blockchain technology are the two cores of the smart revolution, respectively, and can also be considered the two sides of the smart revolution. The union of intelligence technology and block technology can help national governance modernization. Innovative technology can better solve the efficiency problem in national governance modernization. The crucial challenges faced by human society are resource scarcity and production inefficiency. Resource scarcity is an essential factor that generates fierce competition and conflicts in human society. However, the productivity revolution can significantly enrich the means of production and the goods of life, thus profoundly impacting the formation of a productive society in the future. The intelligent revolution, while enriching the means of production, can give people more time to think about more complex problems and engage in more meaningful and innovative activities, which is getting closer to the state of the "association of free men" in Marx's sense.

However, it is essential to note that the biggest problem with intelligence technology is that it can lead to privacy and security issues. These issues then need to be addressed by blockchain technology. Blockchain technology consists of four key technologies: distributed ledger, peer-to-peer transmission, consensus mechanism, and cryptographic methods. Distributed ledgers can better solve the security problem. Since the data is stored in different blocks, this multi-centralized storage method can effectively solve the security problem of a single ledger. On the other hand, the cryptographic approach can effectively deal with the privacy issues that people are most concerned about. The data exchange between people and between blocks through cryptographic encryption algorithms can enhance efficiency while safeguarding privacy, which makes blockchain technology the other side of the smart revolution.

Overall, the key to the modernization of national governance is firstly the productivity revolution with intelligence technology as the core, and the other key is the revolution of production relations with block technology as the core. Combining these two technologies will drive the modernization of governance in developing countries more quickly. On the one hand, it will enable developing countries to achieve leapfrog improvements in productivity. On the other hand, the transformation of production relations can help developing countries solve the long-standing cultural and talent dilemmas they face. Moreover, the integrated role of these two technologies can provide vital support for the modernization of national governance in China and provide new options and references for the development of other developing countries.

After several years of promotion, intelligence technology has been more widely recognized and accepted in China. However, as a new thing, blockchain technology still has many barriers to understanding among the public. It is a crucial difficulty facing the promotion and development of blockchain technology in the future. Therefore, how to better spread, develop, and apply blockchain technology in the future will become an essential part of the future smart revolution.

China may need to make more extraordinary efforts in the development of blockchain technology in the following areas:

First, we should form a breakthrough in the core technology of blockchain. At present, the fundamental technologies of blockchain are still centered on Western technologies. For example, some mature applications in domestic blockchain are mainly built on two major technologies, Ethereum and Hyperledger, both of which are built on foreign platforms and thus may face many risks in the long run. In this sense, developing China's indigenous blockchain core technology is crucial.

Second, the central bank should accelerate the issuance of a national digital currency. If there is a lack of currency support, the application of blockchain technology will lack a focal point. Therefore, the issuance of national digital currency is crucial. Issuing national digital currency can reduce financial crimes and fight corruption. At the same time, it can also help implement blockchain projects that are being vigorously promoted worldwide.

Third, we should be alert to the investment risks of blockchain projects. The application of blockchain technology should be more prominent in problem-solving. Therefore, in promoting blockchain technology, it is necessary to avoid blindly following the trend or even "newsjacking" in various places. We should vigorously promote original projects based on blockchain around the world. At the same time, we should be wary of some unscrupulous elements using blockchain technology in disguise to carry out fraudulent activities in the name of financial innovation, forming a financial risk similar to P2P. Once such fraudulent incidents are frequent, blockchain technology's promotion and long-term development would be seriously hindered.

Contents

1 The Intelligent Revolution and The Global Game Change 1
1.1 The First Two Industrial Revolutions and The World Political Landscape 2
1.2 The Third Industrial Revolution: U.S.-Soviet Cold War and U.S.-Japan Competition 5
1.3 The Fourth Industrial Revolution and the Strategic Conflict Between China and the United States 9
1.4 The Global Game Change in the Context of the Intelligent Revolution 16
1.5 China's Grand Strategy in the Global Game Change 18
1.6 Comparison of the Advantages of China and the United States in the Field of Artificial Intelligence 21
1.7 Eschatology and the Culture of Superman: America's "Achilles Heel" 28
References 33

2 Intelligent Risk Governance and Intelligent Society Building 37
2.1 Taming the "Seviathan" of Artificial Intelligence 37
2.2 Risk Governance and Regulation of Artificial Intelligence 42
2.3 Legal and Ethical Developments in Automated Driving Technology 46
2.4 The Three Major Components of an Intelligent Society 48
References 51

3 The Future of Global Governance of Artificial Intelligence 53
3.1 The Current Structural Dilemma of Global Governance 54
3.2 Analyzing the Hypothesis: Can AI Change the Two Crux Logics of Global Governance 58
3.3 Reinvention or Dissolution? The Role of Artificial Intelligence on the Logic of Hegemonic 61
3.4 De-escalation or Aggravation? The Role of Artificial Intelligence on the Logic of Conflict 72

3.5 Global Good Intelligence: An Effort to Eliminate the Logic of Hegemonic ... 75
3.6 Global Collective Intelligence: An Effort to Eliminate the Logic of Conflict ... 83
3.7 Concluding Remarks ... 89
References ... 91

4 Blockchain and the Transformation of Global Economic Governance ... 97
4.1 The Evolution of the Bretton Woods System and the Expansion of Dollar Hegemony ... 98
4.2 Transformation Scenario: Traditional Centrality and Realistic Polycentricity ... 108
4.3 Technological Features of Blockchain and Its Significance for Global Economic Governance Transformation ... 116
4.4 The Idea of a Supersovereign Currency: The Limits of SDR and the Possibility of E-SDR ... 128
4.5 A New Framework for Global Economic Governance Under E-SDR ... 140
4.6 Conclusion ... 150
References ... 151

5 Blockchain, Global Governance, and China's Grand Strategy ... 161
5.1 The Future of Blockchain and the Intelligent Revolution ... 161
5.2 Blockchain's Revolutionary Implications for the Transformation of the Global Governance System ... 164
5.3 Global Digital Currency Competition and China's Grand Strategy ... 168
References ... 171

6 Intelligent Civilization and Intelligent Social Science ... 173
6.1 The Necessity, Ontology and Methodology of Intelligent Social Science ... 173
6.2 Disciplinary Composition, Issues, and Construction Path of Intelligent Social Science ... 177
6.3 The New Liberal Arts: A New Direction for Liberal Arts Education in the New Era ... 179

Postscript ... 183

Chapter 1
The Intelligent Revolution and The Global Game Change

The World War I and World War II were closely related to the competition for scientific and technological discourse between the leading nations of the first two industrial revolutions. The Third Industrial Revolution, on the other hand, became the technological and contemporary background of the U.S.-Soviet Cold War and the U.S.-Japan rivalry. We can compare the development differences between Latin American and East Asian countries by comparing their different levels of participation in the Third Industrial Revolution. The friction between China and the United States in recent years in areas such as trade can also be analyzed from the perspective of the competition for dominance of the Fourth Industrial Revolution. Artificial intelligence is the leading technology in the Fourth Industrial Revolution. China has developed certain advantages in data, market and governance, but at the same time there are large gaps in intelligent hardware, algorithmic framework, and original innovation. It is suggested that, the U.S. should abandon its current narrow Cold War mindset, instead it need to view the competition and cooperation among countries on AI development from the perspective of the overall interests of human society. While the U.S. has adopted a technology blocking strategy against China, China should respond with a more flexible and competitive cooperation approach. The accelerated development of AI will confront human society with a series of highly challenging issues in the future, and thus requires countries around the world to join together to establish rules and initiatives in the area of AI global governance to safeguard the healthy development of AI.

Artificial intelligence is currently setting off a profound revolution with new historical characteristics around the world. Undoubtedly, this revolution will have an unprecedented impact on human society. The United States has super power in the development of AI technology, while other Western countries may be reduced to the vassal of the United States in the AI era. In the development of artificial intelligence technology, the United States boasted "three wholes" and "three interactions". In contrast, China also has certain competitive advantages, which are mainly reflected in the large and active market and a high level of research team. It should be noted

Q. Gao and J. Zhang, *Artificial Intelligence Governance and the Blockchain Revolution*, Artificial Intelligence and the Rule of Law,
https://doi.org/10.1007/978-981-99-9211-9_1

that the U.S. hegemony has an "Achilles' heel", that is, its AI concept is mainly based on the "eschatology" of Christian culture and the culture of Superman. This concept injects a certain uncertainty into the future development of AI. However, the "symbiosis theory" and humanistic culture developed from the traditional Chinese culture can point to a new direction for the development of AI.

1.1 The First Two Industrial Revolutions and The World Political Landscape

A century earlier, Karl Marx pointed out that science and technology is an important component of the productive forces: "As in the case of the transformation of value into capital, in the further development of capital we see that, on the one hand, capital is based on the existing historical development of productivity as a prerequisite—in these productivity also include science—and on the other hand, capital drives and promotes the productivity forward."[1] Marx viewed science and technology as factors of production as important as capital and labor, etc. In contemporary Chinese studies of international politics, there are relatively few theories that comprehensively analyze international political changes from the perspective of science and technology.[2] This section tries to examine the changes in the international political and economic landscape from the perspective of the Four Industrial Revolutions and, on this basis, explore the impact of artificial intelligence on the current international political and economic landscape.

The First Industrial Revolution was centered on the steam engine, and the leading country was the United Kingdom.[3] The steam engine had been around before the Industrial Revolution. Before James Watt, Thomas Newcomen invented the atmospheric pressure steam engine, but at this time the steam engine was not used in large-scale industrial production.[4] After Watt improved the steam engine, it was used for textiles and then extended to various fields such as porcelain, and greatly promoted the development of handicrafts in England. Since then, England had become the world's factory.

The Second Industrial Revolution was centered on electricity and the internal combustion engine, the leading countries were Germany and the United States. The reason why Germany dominated the Second Industrial Revolution are exemplified by the following. Several of the most significant inventions in the electrical industry were made by the Germans. For example, Ernst Werner von Siemens invented the self-excited DC generator. Generators had been invented before Siemens, but the Siemens generator was the epoch-making equivalent, like as Watt's steam engine.[5] The tram

[1] Marx [1].

[2] The most important works in this field in China is Qixuan [2].

[3] Stavrianos [3], pp. 485–488.

[4] Diamond [4].

[5] Yucang [5], p. 316.

was also invented by Siemens.[6] The core internal combustion engine was first started to be applied by Germans Karl Friedrich Benz and Gottlieb Wilhelm Daimler.[7] These gave Germany a dominant position in the Second Industrial Revolution. Besides, the United States also had dominant position in the Second Industrial Revolution. For example, important inventions in the electric revolution, such as the mass application of electric lights, were associated with the American Thomas Alva Edison.[8] The American Nikola Tesla developed the two-phase asynchronous motor in 1886 and built a 50-kilovolt high-voltage electric line in the Mississippi River basin in 1901.[9] The production model of the flow line was first applied to the automobile industry by the American Henry Ford. In addition, the Wright brothers invented the airplane in the United States, thus creating the aviation industry.[10] Therefore, the most important leading player in the Second Industrial Revolution was Germany, followed by the United States.

However, why did Britain's previous technological superiority not lead Britain to become the leading player of the Second Industrial Revolution? The author believes that there are several main reasons.

First, Britain's domestic market was relatively small, so the wealth that Britain gained through trade was reinvested back into the colonies. This point can explain why industrial production in Britain did not expand any further. Compared to the First Industrial Revolution, the factories in the Second Industrial Revolution were larger and required more capital, so banking would have played an important role in it. But instead of investing money in the expansion of domestic industry, the British banking industry invested heavily in overseas colonial operations, where British capital could make more money on the promotion of overseas colonial operations. This gave rise to a financial model in England: on the London stock exchange, British brokers packaged overseas colony projects into a beautiful product, listed it on the London stock exchange to raise money, and sold it off after making a profit.[11]

Second, the British government became increasingly conservative in its knowledge of cutting-edge technology and related policies. At that time, Britain had already used steam engines for buses and other fields, but the Motor Vehicle Act largely prevented the development of the automobile industry. In 1865, the British Parliament passed the Motor Vehicle Act. The Act stipulated that every motor vehicle on the road must be driven by at least three people, one of whom must be on foot 50 m away from the vehicle to guide it, at the same time, this person also needs to use a red flag to continuously sway to clear the way for the motor vehicle, and the speed must not exceed 4 miles per hour (6.4 km per hour).[12]This Act was later ridiculed as

[6] Zhenhuan [6], pp. 213–214.

[7] Yucang [5], p. 321.

[8] Basara [7].

[9] Yucang [5], pp. 316–317.

[10] Zhenhuan [6], pp. 294–295, 320–322.

[11] Clapham [8].

[12] Britain [9].

the "Red Flag Act". This Act was the result of careful planning by carriage manufacturers and operators, who had vested interests. It was not until 1895, 30 years later, that the Act was repealed. Since the "carriage group" represented the backward productivity, Britain lost the opportunity to become a major automobile country. At the same time, the automobile industry was rapidly emerging in Germany and the United States.[13]

Thus, by the beginning of the twentieth century, Germany and the United States were rapidly taking off in the Second Industrial Revolution, while Britain, despite having a larger economic volume, was no longer technologically advanced during the Second Industrial Revolution. This created a great paradox: the older Britain had a larger economic volume and more colonies, while the emerging Germany, despite being technologically advanced, had fewer colonies. At the same time, Germany challenged Britain in several technological fields, which Britain could not tolerate. This conflict eventually erupted in the form of World War I, which was fought mainly between Britain and Germany. Therefore, this World War can also be seen as a conflict between established and emerging countries in competition for the discourse power in the technological revolution. In international politics, Germany behaved more radically, while the United States adopted an isolationist attitude and showed indifference to the situation on the European continent, so the United States did not have a direct conflict with Britain.[14]

The Second World War can be seen as an extension of the outcome of the First World War. The close relationship between war and technology can be supported by another example. If the German chemist Fritz Haber had not succeeded in industrializing the production of ammonia and using it in explosives, the First World War might have ended around 1914.[15] Of course, the First World War ended in defeat for Germany. Despite the strategic defeat, German industry did not suffer a major blow and, to some extent, developed after the end of the First World War with the support of American capital, so there was a huge contrast between the outcome of the defeat and external constraints such as war reparations and the rise of Germany's own economic power. It was precisely due to the support of giants such as Krupp that Adolf Hitler sought to change the reparations pattern after World War I. In a sense, Hitler's aggressive behavior thereafter was actually opening up new living space for these big businesses, which also played a huge role in Germany's foreign wars.[16]

After the Meiji Restoration in Japan, Japanese companies developed rapidly and caught up with the Second Industrial Revolution.[17] This was an important reason for Japan's dominance in a series of wars, including the Sino-Japanese and Russo-Japanese wars. Some well-known Japanese companies such as Panasonic, Nikon, Mitsubishi and Toyota formed important support for Japan's aggressive activities in the Second World War. For example, Mitsubishi was the main manufacturer of

[13] Lay [10].

[14] Stavrianos [13], pp. 639–659.

[15] Zhenhuan [6], pp. 298–300.

[16] Jianhong [11].

[17] Dongliang [12].

weapons for the Japanese army. The battleship Musashi, the Zero fighter, and various infantry fighting vehicles, heavy tanks, and radars were all produced by Mitsubishi. Panasonic produced almost all of the Japanese military communications equipment and components, Nikon produced military optical instruments, and Toyota produced a large number of military trucks for the Japanese army.[18]

1.2 The Third Industrial Revolution: U.S.-Soviet Cold War and U.S.-Japan Competition

The Third Industrial Revolution is the information revolution, which was accomplished mainly under the leadership of the United States. The Third Industrial Revolution can be divided into the following four stages: the development of the first to the fourth generation of computers, the emergence of personal computers, the emergence of the Internet, and the emergence of smartphones and mobile Internet. These four stages were basically led by American companies.[19]

As mentioned earlier, although the United States has participated in and led the Second Industrial Revolution, however, the advantage was not obvious. In the First and Second World Wars, Germany's strength was greatly depleted, while countries such as the old Britain also suffered attrition, while the United States fought non-domestically and for both times joined the World War midway, which allowed the United States to look on at associated power's trouble with indifference and take advantage of the opportunity to catch up. In the First and Second World War, the United States directly from the war and gained huge benefits, arms trading stimulated the recovery of American industry after the Great Depression. At the same time, the influx of European talent (especially German talent) into the United States during World War II became an important supporting factor for U.S. dominance in the Third Industrial Revolution.[20]

In the Third Industrial Revolution, the greatest challenge to the United States was the Soviet Union. In the early years of computer development, the Soviet Union also built up an information industry of some size.[21] But in the middle and late stages, the Soviet Union lost this competition, and one of the most fundamental reasons for this was that the Soviet Union used a national system in promoting technological progress without effectively translating its military advantages into civilian products, which made Soviet technological progress unsustainable. At the beginning of the Soviet-American competition, the Soviet Union had a tremendous technological advantage, which was reflected in the hydrogen bomb explosion and the space program. At the

---

18 Hongsheng [13].

19 For more information on computers, please refer to Zhenhuan [6], pp. 377–380. For the parts of the internet and mobile internet, see Zhenhuan [6], pp. 403–405. Smart phones can be found in Jianhua and Yiyun [14].

20 Maoxin [15].

21 Goodman [16].

same time, the Soviet Union made great achievements in engineering, mathematics, and physics, which provided a good foundation for technological innovation. The TRIZ system also provided actionable solutions for Soviet innovation.

However, the biggest problem in the Soviet Union was that progress in the entire technological sphere was driven by state power, and the role of the private sector was minimal. More importantly, under the system of planned economy, innovation was inhibited. Since many technologies were not translated to civilian use, they did not work as well as they should have.[22] At the same time, the U.S. engaged in a bait and trap strategy against the Soviet Union. The United States suddenly adopted a strategy of de-escalation in the Soviet competition and offered electronic components at prices lower than the Soviet Union's own R&D costs. The Soviet Union felt that such products were both cheap and useful, and then gradually abandoned the production of its own components and bought them from the West step by step. Until the 1970s and 1980s, the Soviet computer industry was increasingly vulnerable in the competition, until finally eliminated.[23]

Japan began to promote the development of the chip industry in the 1970s and gradually became the most important competitor of the United States. On the one hand, Japan's rise in the chip industry mainly relied the leadership of the Ministry of International Trade and Industry. Through industrial policy, Ministry of International Trade and Industry (MITI) carried out industrial integration and united some competitive companies in the chip field to form industrial alliances to jointly cope with industrial competition from the international arena.[24] On the other hand, Japanese entrepreneurs responded to the call of the national industrial policy, and entrepreneurs and researchers carried out a large number of technology introduction and technological innovation.[25] As a result, by the mid-1980s, Japan's semiconductors had overwhelmingly surpassed those of the United States, which is an important reason why the United States signed a semiconductor agreement with Japan.

The United States initially saw Japan as an overseas factory in the U.S. division of labor in the world economy, and therefore did not take much precaution against Japan's semiconductor industry. But when Japan's chip industry as a whole rose, the U.S. used very tough tactics to deal with Japan. The Japan-U.S. Semiconductor Agreement is an important example. Not only that, the United States also supported other economies to suppress Japan's related industries, such as support for South Korea's semiconductor industry. In order to reduce the cost of production of their own enterprises, the United States has adopted a new joint approach, that is, U.S. companies mainly engage in research and development and standards setting, while allowing South Korea companies as foundries for chip processing. On the one hand, this combination allows the U.S. to reduce its investment in production and focus on R&D. On the other hand, the U.S. occupies the upper reaches of the entire industry chain and can define the entire industry, thus gaining the largest part of the global

[22] Weiying and Bin [17].

[23] Adirim [18].

[24] Fei [19].

[25] Woo-Cumings [20].

value chain. By the end of the last century, Japan's semiconductor industry began to decline, and today there are no Japanese companies among the top 10 global semiconductor manufacturers. In 2017, Toshiba was still among the top 10 global semiconductor manufacturers, and by 2018, Toshiba was also acquired by a U.S. company.

In addition to the U.S. strategic strike, the decline of the Japanese chip industry also has the following three reasons. First, the situation in Japan is very similar to that of the Britain. Japan's relatively small domestic market size has led to capital outflows and investments around the world.[26] Japan has a lot of assets overseas and richly profitable investments, but the overall trend of Japan's domestic enterprises is decline. Second, after Japan experienced the real estate bubble in the 1980s and 1990s, the country as a whole adopted a relatively lenient education policy. Since then, Japanese nationals have entered Loose Generation. The mental state of the entire Japanese nation began to fall into a slump.[27] Especially the youth, who should be energetic, lacked hope and longing for the future. Kenichi Ohmae described such a scenario in his book "Low Desire Society".[28] Third, along with the aging of Japan's population, it is difficult for more money to be spent on reinvestment and R&D after the huge public spending of the limited government coffers.[29] This has largely limited Japan's industrial upgrading, which has led to the successive withdrawal of some of Japan's dominant industries from the market, such as LCD panels and semiconductors.

A classic proposition in comparative political studies is the comparison of the development of the United States and Argentina. Both the United States and Argentina were considered to be the most promising countries in the early twentieth century, but it turned out that the United States became the most powerful country of the twentieth century, while Argentina fell into a middle-income trap. Another classic pair of case comparisons in comparative political studies is Latin America and East Asia. After the end of World War II, Latin America and East Asia had relatively similar economic levels, but half a century later, development outcomes diverged considerably. By the end of the twentieth century, many East Asian countries and regions had completed industrialization and had become newly industrialized countries, while most Latin American countries remained underdeveloped.

In the middle and late last century, Latin American scholars proposed the dependency theory. The ideological source of the dependency theory is mainly Marx. Using the theory of surplus value, Marx thoroughly analyzed the secret of value is that capitalists exploit workers by exploiting surplus value: "Although only a part of the worker's daily labor is paid and another part is unpaid; and although it is this unpaid or surplus labor that forms the basis for the production of surplus value or profit, it appears on the surface as if all labor were paid labor."[30] Lenin

[26] Junwen [21].

[27] Yimin [22].

[28] Ohmae [23].

[29] Wenqing and Rundong [24].

[30] Marx [25].

put this Marxian theory into an international perspective and developed the theory of imperialism. Lenin believed that imperialist countries plundered other countries by exploiting and extracting surplus value from other countries as well as by state monopolies. Lenin wrote: "Imperialism is the monopolistic stage of capitalism. Such a definition would cover the main points, because, on the one hand, financial capital is the banking capital of the few monopolies of the largest banks fused with the capital of the monopolistic alliance of industrialists; on the other hand, the carving up of the world is the transition from the colonial policy of unhindered access to regions not occupied by any of the great capitalist powers to the colonial policy of monopolistic possession of the already carved up world territory."[31] On the basis of Marx and Lenin's theories, Latin American scholars further advanced and proposed a new dependency theory. It argues that Latin America's underdevelopment is caused by Latin America's dependence on the United States, and this dependence is itself a manifestation of inequality.[32] The development of the dependency theory theoretically guided the left-wing movement in Latin America at the time, and a large number of left-wing governments came to power in Latin America during this period. The import substitution strategy, which was accompanied by the dependency theory, was adopted by the vast majority of Latin American countries, i.e., to reduce their dependence on others by building a more comprehensive domestic industry centered on their own markets.[33] From the current results, it can be seen that this strategy in Latin America has almost been disrupted, while the debt crisis in Latin America has become even more severe. When the economic problems are not solved, there will be military intervention.[34] As a result, Latin America has not done particularly well for a longer period of time. In contrast, East Asian countries have made rapid progress, including the "Four Little Dragons" and the "Four Little Tigers", which are examples of more rapid development.[35]

Overall, the difference in development outcomes between Latin America and East Asia can be understood in terms of technological participation in the Third Industrial Revolution. East Asian countries joined the great wave of the Third Industrial Revolution by participating in the world division of labor. South Korea participated fully in the Third Industrial Revolution, developing information industry giants such as Samsung and Hynix in cooperation with the United States. Samsung played the role of a foundry in its early development, but saw further growth when cell phones became an important consumer product.[36] Since the information industry, especially the chip manufacturing industry, requires large-scale industrial agglomeration and financial support, East Asian governments have played an important role in the growth of the information industry. In the academic field, this development model in

[31] Ref. [26].

[32] Dos-Santos [27].

[33] Baer [28].

[34] Huntington [29].

[35] Lin et al. [30].

[36] Nenni and Dingee [31].

East Asia is called developmental state or developmental government. These countries and governments first make comparative clear plans, and at the same time gather large amounts of capital through industrial investment and other means for the manufacturing of industries such as chips.[37]

In contrast, Latin America did not participate sufficiently in the Third Industrial Revolution to form globally competitive enterprises. In the explanation of Latin America's failure, the West mostly believes that it is the workers' movement in Latin America as well as the corruption problem that led to the failure of Latin America's economic transformation.[38] These factors all pose certain obstacles to the Latin American economy. However, the author believes that the key to the difficulty of sustained long-term economic growth in the Latin American region lies in the dependence theory, which has resulted in the Latin American region not participating in the world economic division of labor for a long time. It deprived the Latin American region of the opportunity to fully participate in the Third Industrial Revolution. Therefore, emerging countries need to rise from within the Western capital system. Because of the dominance of new disruptive technologies held by Western capitalism, the first step to winning victory over imperialism is to win technological competition with it. The Soviet Union, because of its relatively strong scientific and technological base, could have competed with it in science and technology outside the Western capitalist system. However, the Soviet Union's competition also ended in failure. For developing countries, the best way is to develop scientific and technological power from within capitalism and then compete with them internally.

1.3 The Fourth Industrial Revolution and the Strategic Conflict Between China and the United States

This series of technologies associated with the Fourth Industrial Revolution includes artificial intelligence, Internet of Things, blockchain, virtual reality, brain-like computing, gene technology and so on. One of the most critical technologies is artificial intelligence technology. On the one hand, the Fourth Industrial Revolution has a disruptive impact.[39] On the other hand, the Fourth Industrial Revolution is still highly dependent on the information industry, which is the most crucial in the Third Industrial Revolution, so the Fourth Industrial Revolution can also be seen as a further upgrade of the Third Industrial Revolution. Therefore, the United States, which had the dominant power in the Third Industrial Revolution, still holds a huge advantageous position in the Fourth Industrial Revolution. Whether in the talent reserves, intelligent hardware, algorithm framework, and some specific AI application scenarios, the United States undoubtedly has a huge advantage (Table 1.1).

[37] Jun [32].

[38] Lipset and Lenz [33].

[39] Schwab [34].

Table 1.1 Historical comparison of the four industrial revolutions

	Core technologies	Dominant countries	Major disputes	Results
The first industrial revolution	Steam engine	United Kingdom		Britain becomes the world's factory
The second industrial revolution	Electrical and internal combustion engines	Germany, USA	World War I, World War II	The outbreak of two wars
The third industrial revolution	Information technology	United States	U.S.-Soviet Cold War, U.S.-Japan rivalry	The U.S. has a potential advantage in the fourth industrial revolution
The fourth industrial revolution	Artificial intelligence related technologies	USA? China?	U.S.-China strategic conflict	

China is more often seen by the West as an "intruder" in this field. From the perspective of China's own development, China has seized the opportunity of its time well. After the founding of the country, China's large-scale industrialization laid a solid foundation for the construction of heavy industry. The greatest significance of the reform and opening up is the activation of the private economy. The awakening and revitalization of the private economy has provided important assistance for China to catch up with the process of the Third Industrial Revolution. For example, Huawei entered the communication industry at the beginning of its startup in 1987, rooted in the information base on by the Third Industrial Revolution. Again, as in, at the time of Ali's founding, Ali was no worse than the U.S. companies in terms of its start-up time, despite the emergence of e-commerce companies such as eBay. At the same time, Ali's growth has been very rapid. Ali has developed faster than its foreign counterparts in the entire e-commerce ecosystem, especially in the development of Internet financial products such as Alipay.[40] Internet companies in the U.S. experienced a speculative bubble from 1995 to 2001, and many technology companies entered a cold winter. During the same period, Chinese Internet companies only began to rise. Despite their late start, they also avoided the Internet winter at that time to a certain extent. At the same time, the development history of U.S. Internet companies also provides more experience and lessons for Chinese Internet companies. For example, Tencent's earliest software, OICQ (the predecessor of QQ), was built on the basis of imitation of ICQ. But based on the rapid development of China's domestic Internet, Tencent established good roots in the domestic Internet industry right from the start.[41] WeChat, a newly developed product, has enabled

[40] Erisman [35].

[41] Xiaobo [36].

Tencent to take a step further in the mobile Internet industry. Currently, the WeChat platform is at the forefront of global social media, breaking through the limitations of traditional chat tools and moving toward a holistic platform.

In China, Baidu's growth was made possible by Google's exit. Google chose to withdraw from the Chinese market in 2010 due to a series of policies such as concerns about regulation in China. Google's exit provided an opportunity for Baidu's rapid development. Since then, Baidu has established a search business-centric development model based on the Chinese market and has made great progress. Currently, based on its large data stock of maps and search information, as well as its advantages in natural language processing, Baidu has put forward its artificial intelligence development strategy earlier and has also become the leading artificial intelligence company in China in line with the trend.

In fact, in the field of artificial intelligence, in addition to the currently well-known BAT, China has a large number of increasingly leading companies in the international arena, such as IFLYTEK, SenseTime, YITU, MEGVII, AISpeech, etc.. At present, China's artificial intelligence companies have certain advantages in the field of speech recognition, image recognition and other applications.

Overall, China has three main advantages as follows: First, the advantage of massive data. Because of the high penetration rate of smart devices in China and the large population base, there are more smart device users in China. The current development of artificial intelligence mainly relies on data-driven deep learning and other algorithms, with more emphasis on data-based learning and training. In this way, the data advantage that China has can be transformed into a technological advantage. Second, the advantage of a unified and active market. The market is an important foundation for the rise of Chinese Internet companies. Because China's market is large, vibrant, unified, it is easy for AI companies to grow rapidly on a level playing field as long as they have good products. This advantage is unique in the world. Third, the advantage of strong government support. Before the AI revolution, China was already promoting the "Internet+" strategy. Although the "Internet+" strategy is more concerned with mobile Internet applications, it has already laid a good foundation for the development of AI. Under the influence of the "Internet+" strategy, a number of globally competitive companies such as DiDi have emerged.[42] On this basis, the intelligent transformation of Chinese enterprises can be more convenient. For example, DiDi can move into the driverless field more easily in the future. In addition, for AI, the Chinese government has now designed a series of plans that are expected to promote the overall development of AI.

China's advantages have been reflected in the new round of 5G implementation and related applications. Companies such as Huawei, ZTE, and China Mobile have made tremendous efforts in the laying of 5G and the competition for discourse power, which is an important reason for the US to impose sanctions on these companies. These events also reflect the influence of Chinese companies in the 5G space from the side. Huawei not only has an advantage in communication infrastructure, but is also gradually developing in smart devices such as cell phones, which includes various aspects such as baseband chips and the production of high-end cell phones. Huawei's

[42] Jiangan et al. [37].

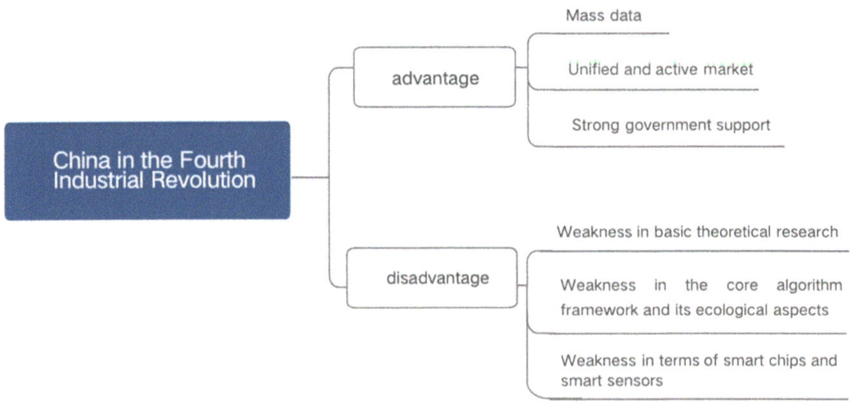

Fig. 1.1 China's advantages and disadvantages in the fourth industrial revolution

Hense Semiconductor has also become the most competitive semiconductor company in China. In addition, in the field of chips for artificial intelligence, companies such as Cambrian and Horizon have also performed very well.

However, China still has some obvious shortcomings in artificial intelligence. First, China is still relatively weak in basic theoretical research. There is still much room for improvement in China's original research on AI-related theories, methods, and systems. Second, in terms of core algorithm frameworks and their ecology, China is still relatively weak. At present, mainstream frameworks such as TensorFlow, SystemML, Caffe, Torch are dominated by the U.S. companies or research institutions, while China's domestic algorithmic frameworks are relatively few, and the ecology is not complete, and still lack the ability to compete with foreign mainstream frameworks. Third, in terms of smart hardware, China has made great progress in some application-based chips. For example, HUAWEI Kirin series, Cambrian's NPU, Horizon's BPU, WESTWELL's deepsouth and deepwell, Alibaba DAMO Academy's Ali-NPU under research, and Unisound's UniOne and so on. However, in terms of basic chips, globally dominant products such as NVIDIA's GPU, Intel's NNP and VPU, Google's TPU,, IBM's TrueNorth, ARM's DynamIQ, etc. are almost all occupied by Western companies. In terms of intelligent sensors, although China has made some progress in a relatively single field, such as Collihigh's force sensors and Goodix's fingerprint sensors, there is still a big gap between China's sensors and other giants's sensors in terms of overall layout and so on (Fig. 1.1).

In viewing China's technological development, the United States still adopts a traditional Cold War mindset, or hegemonic thinking. On the top of three pillars of science and technology, finance, and military, the U.S. sustains global hegemony.[43] Through its monopoly of high-end technologies and the continuous expansion of emerging technologies, the U.S. occupies the discourse of emerging technologies, thus preventing its rivals from entering the field of emerging technologies. At the same time, through discursive hegemony in the field of technology, the U.S. achieves

[43] Santos [38].

the definition of new technologies while taking away the most valuable part of the global value chain. In the financial sphere, the U.S. provides the world with dollars and levies a "minting tax" on the world. Military power is the guarantee of U.S. political objectives and the basis for U.S. technological containment of other countries. For example, the U.S. containment of Japan's technological rise is tied to military considerations. At the same time, these three pillars intertwine and influence each other. The development of U.S. high technology is closely linked to the Defense Advanced Research Projects Agency. The research and development of many emerging technologies is often led and promoted by the US military first, and then shifted to civilian research and development and promotion to address cost issues. At the same time, this transformation model also stimulates the vitality of civilian and market-based science and technology innovation. Meanwhile, the U.S. has also used its financial position and stock exchange market to raise funds for high-tech research and development and for the transformation of scientific and technological achievements.

When the U.S. hegemony is challenged, the U.S. will use all kinds of power to protect its hegemony, that is, to use technology, finance or military to counterattack. The outward appearance of U.S.-China friction is trade friction, but its deeper essence is technological competition. The United States has adopted a series of measures and instruments related to the technology blockade to impose restrictions on China. For example, restrictions on Chinese companies acquiring U.S. high-tech companies, restrictions on Chinese students studying abroad, and restrictions on cooperation between U.S. research institutions and advanced Chinese companies, etc.

This narrow Cold War thinking of the United States is becoming more and more anachronistic in the era of AI. In the age of intelligence, a holistic thinking on the future direction of AI needs to be formed at the global community level. The reasons are as follows:

First, AI may exacerbate the advent of a risk society. For example, AI may be used in some black industries. In a sense, AI is used in the black industry faster than the white industry. Under the temptation of huge economic benefits, the practitioners of black industries can use AI in some non-compliant fields more recklessly in the absence of legal and moral constraints. For instance, voice imitation, intelligent face replacement and other AI-related forgery technologies are likely to be exploited by fraudsters, which may give rise to more serious political and social risks.

Second, the use of AI in the military may increase military competition among countries. When the U.S. was developing AI, it was initially driven by the military. The U.S. military wants to use AI on the battlefield and has already had relevant deployments, such as the large number of drones being developed and applied. The initial consideration of the U.S. military was that these intelligent devices, when used on the battlefield, could reduce casualties among U.S. soldiers, but this could increase the gap between the U.S. military and the military forces of other countries even more. Meanwhile, military AI can also create huge ethical problems. Decision-making by machines in military operations could become an excuse for the U.S. military to shirk its responsibilities. For example, after an attack against civilians in a drone military operation, the U.S. military evades responsibility by attributing responsibility for the decision to the machine.

Third, as a disruptive technology, the impact of AI on human society will spill over to other countries. For example, the risk of unemployment caused by technological advances in AI is likely to have a global impact. If the risk of widespread unemployment spreads around the world, it could lead to serious social problems.[44] At the same time, the risk of unemployment would also exacerbate the wave of anti-immigration, and this trend is already occurring in Europe and the United States. Therefore, sovereign countries need to unite to think about these issues and jointly plan for the overall development of artificial intelligence.

Fourth, countries should reach a consensus on the research and development of general AI. Most Western countries currently encourage further development in general AI, and the Asilomar principle, for example, does not oppose the development of general AI.[45] But the eventual development of general AI is likely to produce a great challenge to the meaning of human beings. If the development of general AI eventually leads to the loss of meaning for all the values of human existence, then this would be unacceptable to humanity. This requires countries to join together to reach a basic consensus on the direction of the development of general AI.

In the new round of technological revolution, artificial intelligence is both a strategic technology and a key technology leading other technological breakthroughs. The development of artificial intelligence is having an extremely profound impact on economic development and the international political and economic landscape, etc. China is making certain breakthroughs in the development and application of artificial intelligence technologies on the ground. Because of this, the U.S. has taken a series of measures to block China's AI development as well as China's technological progress. These actions include restricting Huawei's layout on 5G applications, exchanges between high-end talents from China and the US, and Chinese students studying in the US. Against this backdrop, China needs to be more determined to make breakthroughs in the development of AI by making breakthroughs in fundamental key technologies. In a series of core areas such as chip manufacturing, core algorithms and operating systems to form China's own advantages. At the same time, the great significance of the development of artificial intelligence is reflected in its deep integration with social and economic development. Artificial intelligence is not only a basic technology, but also an applied technology. The key for artificial intelligence to play a role in scientific and technological change and industrial change is to deeply integrate with primary, secondary and tertiary industries and create products that solve practical pain points.

In short, the technical attributes of AI should be highly correlated with social attributes. Through the industrial cultivation and product application of AI, it will provide comprehensive support for China's scientific and technological progress, social development, and national security. In the development process of AI, core

[44] Qiqi [39].

[45] At the 2017 Beneficial AI Conference in Asilomar, California, several hundred experts in AI and robotics co-signed the Asilomar AI Principles of Artificial Intelligence. Articles 9, 10, 16, 17, 19, and 20 of the principles all refer to and allow for the development of general AI. Also, for future risks of AI, the Principles 21 and 22 advocate self-healing. Future of Life Institute, Asilomar AI Principles, available at https://futureoflife.org/ai-principles/.

technology is the foundation. And in the process of application, producing a series of products that can solve social problems is the key. "China's New Generation Artificial Intelligence Technology Industry Report" by State Council of the People's Republic of China points out that by 2030 China will become the world's major AI innovation center, and the theoretical and technological applications of AI will generally reach a world-leading level. It should be emphasized that the theory here should include not only the scientific theory of AI, but also the related social theory. Artificial intelligence in the landing will generate a series of social issues, such as privacy and security. In addition, the social application of AI is also in a less researched field in terms of morality, ethics, employment and government governance, and there are very few relevant studies and achievements abroad. Since China wants to be at the forefront of AI applications, it is equally important to explore these issues.

At present, great powers have begun to do researches and develop standards and guidelines in the field of artificial intelligence, such as the EU, which released its AI ethics guidelines in April 2019. Some large companies in the United States are also advancing relevant principles or rules, such as the Asilomar Principles. And China's voice in these rule-making is still relatively weak. Technology is the basis for the development of AI, but at the same time the establishment of related ethics, policies and laws is the key to the application of AI to social governance. Moreover, China should represent the position of developing countries in the development of new rules so that it can better compete with developed countries. Meanwhile, the super power like the United States in the development of AI will not disappear in a short period of time. Whether in the core theory of AI, the framework of algorithm application or the ethical rules, the United States has an advantage that is difficult to be shaken by other countries. Therefore, in such a context, competitive cooperation with the U.S. will become very important. Although the U.S. has adopted a technology blocking strategy against China, China should respond with a "roll with the punches" approach. And, in the development of AI, cooperation is also a rigid necessity. Because AI may eventually produce results that are subversive to humanity, there are various dangers associated with the development of general AI, for example. In this sense, nation-states need to join forces to address these risks as well. Chinese President Xi Jinping pointed out that "dealing with the new topics raised by AI in the areas of law, security, employment, ethics and government governance requires deepening cooperation and joint exploration by all countries. China is willing to work with all countries in the field of AI to promote development, protect security, and share results."[46] Therefore, cooperation is not only a passive strategy for China to respond to the U.S. challenge, but should likewise be a proactive strategy for China to adopt. Therefore, China needs to be more proactive in cooperation with the U.S. and other countries under the framework of competitive cooperation to establish a series of international rules and initiatives for the future development of AI, so as to guarantee the healthy development of AI.

[46] Congratulatory Letter from President Xi Jinping to the 2018 World Conference on Artificial Intelligence, September 17, 2018 (in Chinese).

1.4 The Global Game Change in the Context of the Intelligent Revolution

The Global Game Change refers to the change in the distribution of power among major countries in the world under the conditions of the new scientific and technological revolution, and the new imbalance in the power structure of the international system. "Change" is the norm in the game change. "Change" is the essence of traditional Chinese culture. The Book of *Zhou Yi* emphasizes: "If you are poor, you will change, and if you change, you will pass, and if you pass, you will last." (The Book of *Zhou Yi - XiCi* Part 2). "Change" is also the driving force of development. In the the Book of *Tao and Teh,* it is said, "Reversion is the action of Tao." However, the game change means that the whole international landscape is being profoundly adjusted and the new direction of development is not fully formed, in which there is still a lot of uncertainty.

Specifically, the current global game change is mainly reflected in the following aspects.

First, a new shift in leadership in the global system is emerging. The United States is considered to be the leading power in the post-World War II Bretton Woods system. Realism proposes a theory of leadership stability, which argues that the stability of the international system depends on the leadership state providing public goods. However, the current situation is that the United States is abdicating its role as a leadership state. A recent series of actions in the United States have expressed its rejection and instability of the global public order it has participated in establishing, which is commonly referred to as the "retreat" behavior.

Second, profound changes are also taking place within the European countries that were previously involved in global leadership. European integration and the development of the European Union have always been considered one of the greatest processes in human society, namely the advancement of supranational unity without violence but through consultation. The Brexit, however, could be a watershed for the European integration process. In addition, the migration impact faced by Europe and the internal economic sluggishness are uncertainties for its future development.

Third, emerging countries as a group is generally rising, and the G20 is an outward reflection of this rise. However, the development speed and situation faced by emerging countries are vastly different. For example, Brazil is facing economic recession and corruption, and Russia is facing embarrassment in the context of new geopolitics. Therefore, there is still great uncertainty for emerging countries in the future. Moreover, in terms of the international landscape, the global mechanism does not profoundly reflect the rise of these emerging countries.

Fourth, the global southern countries have also occurred differentiation. Some countries in the South are emerging, but the vast majority of the South still faces challenges on all fronts. Many countries in the South still face serious underlying problems such as poverty and underdevelopment.

The current global game change is taking place under the influence of the Fourth Industrial Revolution. What is happening now is the Fourth Industrial Revolution, the intelligent revolution. There are two core technologies in the intelligent revolution: artificial intelligence and blockchain. What artificial intelligence achieves is the intelligence of things, and what blockchain achieves is the intelligence of relationships. Many of the current AI technologies are developed on the basis of the computer revolution, so the US still has a very strong voice in this industrial revolution. Although other countries such as Germany have proposed new concepts such as Industry 4.0, Germany, Japan, Britain, Canada, Israel, etc. are extremely inferior compared to the great strength of the United States in this field.

Current intelligence technologies are mainly built on the basis of big data, or data-driven intelligence. Technologies such as neural networks or deep learning, which have been highly relied upon and promoted by the intelligence revolution in recent years, have to be built on the basis of large amounts of data. The use of existing data to train new models to predict the future has become a major trend in the development of science. In such a context, data becomes the most important asset, and this may be an important advantage for countries with large populations, because data is accrued in human units. In the era of Big Data, data is the new power. By assembling the data and then training the models with techniques such as neural networks, some predictions can be made for the future as well as decision-making references.

Countries with larger populations like China and India have the advantage of data accumulation in a sense, and data will become the most important resource and weapon in the future. The General Data Protection Regulation published by the EU is protecting the personal data of European citizens from being easily used by big giants like Google. This regulation emphasizes that data belongs to the people and encourages European companies to redevelop this data. Because Europe has already felt its technological decline in the era of big data, they do not want to widen this gap. Therefore, Europe has adopted the General Data Protection Regulation to prevent monopolistic exploitation of data by foreign companies on the one hand, and on the other hand, Europe promoted the gradual growth of local companies in data mining and development through the Framework Regulation on the Free Movement of Non-Personal Data.

In a sense, the technologies associated with the Intelligent Revolution have a certain flattening effect, which is reflected in the following two points.

First, the most important feature of smart-related technologies is open source. This has been reflected in the computer era, for example, the computer operating system Linux and the cell phone Android are open source. The current tradition formed in the computer field is that when a major discovery or research progress appears, they tend to publish it on the arXiv website and also upload the code to the github platform to facilitate further development by other researchers. Again, as in, Python has become the most popular programming language among developers, because many developers have developed third-party libraries based on Python, which allows subsequent researchers to call on this foundation. This reflects the characteristic of human society of crowd wisdom, which means that mutual cooperation can produce more results. This concept is similar to crowdfunding, and its role in technology development is growing. A similar development model is used by the well-known Chinese software company "ZBJ.com".

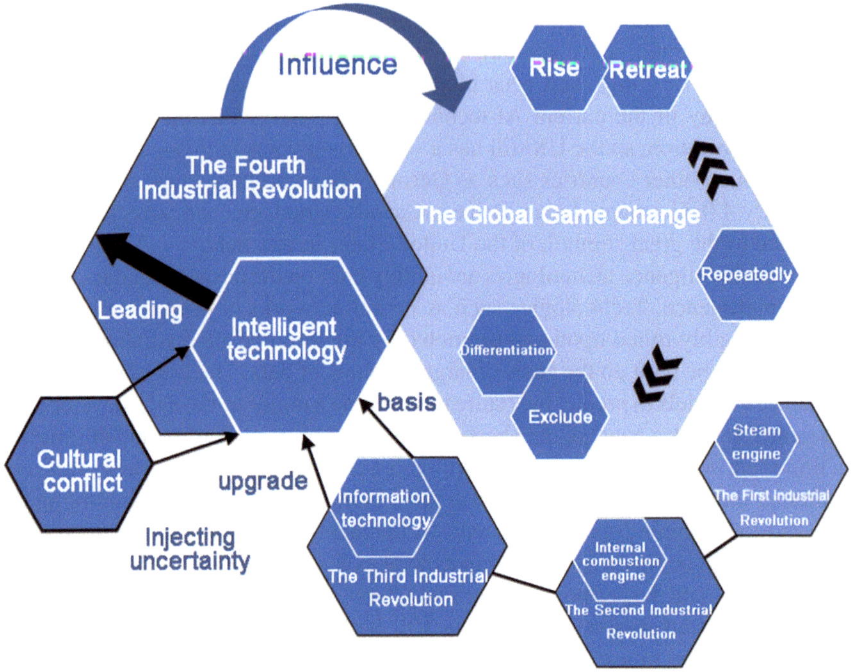

Fig. 1.2 Industrial revolution and global change

Second, the time gap between the leaders and the catchers is narrowing. In previous industrial revolutions, there was a significant time gap between leaders and catchers, for example, when Britain launched the First Opium War, China was still an agricultural country, and the technology gap between China and Britain could be decades or even centuries. But in the era of intelligent revolution, the developed countries with technology release a certain achievement, such as smart phone, driverless or cashless payment, while the developing countries may have similar achievements a year or two later, so such time gap is getting smaller and smaller. This facilitates the learning and catching up of developing countries, and in a sense has a helpful meaning for the disadvantaged countries, while all these features intensify the complexity of the great global game change (Fig. 1.2).

1.5 China's Grand Strategy in the Global Game Change

China has played a very special role in the global game change. Because the old developed countries are losing power and China is gathering new development power on behalf of the developing countries, it is making a huge contribution to the global economic development in the context of the global economic downturn. China is currently unique in that in one respect it has a huge industrial base, and in another it has used its current significant demographic advantages, which are being transformed into new data advantages and Internet application advantages. These advantages are closely related to a number of factors such as China's vast market, relatively intact industrial base and relatively high population quality.

However, in the context of the new technological intelligence revolution, China still has serious shortcomings in certain areas, for example, in terms of primary innovation as well as core technologies, China still has many shortcomings and deficiencies. Therefore China should put more emphasis on the following points in its future development.

First, form a breakthrough in core technology. In the ongoing intelligent revolution, the core technology of almost all important hardware is in the hands of the developed countries in the West, and China is still facing many bottlenecks in the technology of chip manufacturing. It has been until after the ZTE incident broke out that China has attached great importance to the research and development of chip technology. Therefore, chip technology is the core technology that China needs to develop in the future. China has made "Internet+" and artificial intelligence as strategic goals to develop; however, such prosperity needs to be built on the basis of hardware facilities. Without hardware support, software cannot run. Then again, the operating system is also the key to future development. Because the operating system involves the security of the system and a series of network security most core issues, and the current operating system of both computers and cell phones are basically developed by foreign countries. Foreign giants in the operating system with ecological advantages are difficult for domestic companies to catch up in the short term. These require us to carry out long-term strategic reserves and overall research and development.

Second, put more emphasis on education. On the one hand, it is important to develop new strengths in STEM-related education. Since the reform and opening up, China has made remarkable progress in education, but Chinese education has put more emphasis on knowledge learning and neglected the cultivation of students' innovative abilities. And in the context of the current intelligent revolution, the traditional education model may be completely overturned. There is a huge tension between the traditional education model based on knowledge memory and the development needs of the future intelligent society. On the other hand, the cultivation of intelligent talents is the key to the future intelligent revolution. The Chinese attach special importance to education in general and invest more and more in education, such as the investment in the construction of high-level universities in China in recent years, which are favorable conditions. However, how to do some innovative attempts in the existing educational framework is a further question. It can reserve new talents for the future intelligent revolution. For example, talents in artificial intelligence are currently very scarce, and how we can produce enough artificial intelligence talents to meet the needs of our market and social development in a relatively short period of time is a major practical problem. Then again, there is the greatest lack of composite talents who understand both AI technology and at the same time understand application scenarios and specific related knowledge. Although the AI Innovation Action

Plan for Higher Education issued by the Ministry of Education is already emphasizing this point, the cultivation of such composite talents is still a long way to go.

Third, form a new voice in the scientific and technological revolution. The new discourse is manifested both in the scientific and engineering sense, such as the formation of groundbreaking theories in certain scientific fields. But at the same time, it is more important to form a new discourse in terms of standards, security, privacy, social ethics and other social science research results. Because many of the effects of the intelligence revolution on human society are yet to occur, it is difficult to conduct in-depth research on these issues with traditional research methods. As a result, research results in these areas are almost always in short supply and will be in very high demand in the future, so that forward-looking research results in these areas may become the new theoretical high ground in the future. For the first time in two hundred years, China is standing on the starting line of the new intelligence revolution at the same time as the developed Western countries. China has such strength and theoretical basis to do research on related issues. Since the major issues of human society are mainly determined by social concepts, it is especially important to form a new discourse in the relevant social science fields. Of course, this intelligent revolution also involves some relevant research results in other fields, such as life science and material technology. Therefore, the social science fields related to the whole science and technology should all become new fields and new starting points for the development of social science in China in the future.

Fourth, China should take into account the production of global public goods within its capacity. In the current context of global game changes, an important feature is the lack of global public goods. The United States has always been responsible for the production of public goods, but a series of recent "withdrawals" behaviours have left a vacuum of public goods in the international community. All countries have shifted their focus back to within their nation-states while neglecting the international community. In fact, there is a close interaction between the domestic development of countries and the international community. If there is a serious lack of global public goods, then the problems of each country will generate many disputes in the international community, which will eventually return to the domestic level and create a cluster of domestic problems. Therefore, in such a context, the production of global public goods becomes very important. Although China's overall strength is growing, China's power to undertake global public goods production is currently far from sufficient. China is still an emerging economy and faces many challenges in its future development. In this context, China's "the Silk Road Economic Belt and the 21st-Century Maritime Silk Road" initiative and the idea of "a community with a shared future for mankind" are very important, which means that China helps other developing countries as much as it can, but it also depends on the cooperation intentions of other countries. Therefore, "the Belt and Road Initiative" is more of an initiative for the common development of all countries, while "a community with a shared future for mankind" is the ideal state for the future development of the global society as a whole.

The "change" in the global game change is mainly reflected in the change of power distribution, that is, the power distribution of the international community has changed. The developed countries have a tendency to decline, while the emerging countries are rising, which is the "change" in the global game change. At the same time, the global game change also shows that the direction of future development is not clear, and the trend is not yet fully formed. The intelligent revolution which as the fourth industrial revolution, is taking place, and while the United States, as a traditionally developed country, and China, as an emerging country, both have certain advantages, however, there is still uncertainty about the future. The "unchanging" in the global game change is reflected in the global issues. Some important global issues are still prominent, such as climate change, environmental pollution, energy issues, security issues, and so on. The global community still faces great challenges in solving these problems in the future. Although the intelligent revolution has provided mankind with a new possibility to solve these problems, the current state of affairs shows that there is still a long way to go to solve these global problems. We cannot underestimate the difficulties faced in the process of solving these global problems. Overall, the concept of global game change brings us a new way of thinking today: the current global landscape is still full of great variables, and we do not know the exact direction of the future development of the global community. While the "a community with a shared future for mankind" is China's proposal for the future direction of the global community, the global game change tells us that the road toward this goal is still full of thorns and requires the joint efforts of people from different countries and ethnic groups to overcome differences and reach consensus.

1.6 Comparison of the Advantages of China and the United States in the Field of Artificial Intelligence

What is the current pattern in the development of AI in the world? Where does China stand in the current landscape and what role will it assume in the future? First, this section discusses the impact and even reshaping of artificial intelligence on human society; second, this section analyzes the super power of the United States in the era of AI and its impact on the current world landscape; third, it compares and analyzes the advantages of China and the United States in the field of AI; fourth, it analyzes the "Achilles' heel" of the development of AI in the United States, i.e., the "eschatology" and the culture of Superman; finally, the moral position and special mission of China in the AI revolution are explained.

Artificial intelligence is not only a new technology that changes human life, but also an unprecedented social revolution that has a powerful impact on human life. Compared with the three previous industrial revolutions, AI will have a much broader and deeper impact on the future of human life. The first three industrial revolutions powered human life in their own unique ways. For example, the steam power provided by the First Industrial Revolution brought human beings into the steam age; the electricity and internal combustion engine power of the Second Industrial Revolution brought human beings into the electrical age and the internal combustion engine age;

the information power brought by the Third Industrial Revolution brought human beings into the information age. Unlike the first three industrial revolutions, which mainly impacted the primary and secondary industries, the AI revolution will have a disruptive impact on all industries and even reshape certain industries. Specifically, AI will have the following impacts on various industries:

First, the development of artificial intelligence will have a huge impact on traditional industries, and some industries will even disappear under this impact. For example, the development of machine translation will have a huge impact on the translation industry, including written translation, oral interpretation and simultaneous interpretation.[47] It can be expected that machine translation will reach or even exceed the level of human translation in the future. In addition, advances in speech recognition technology have put court clerks and meeting stenographers at great risk of losing their jobs. For example, in February 2017, the Supreme People's Court published "Several Provisions of the Supreme People's Court on Audio and Video Recording of Courtroom Activities" highly affirmed the application of intelligent voice technology in courtrooms and stipulated that system-generated transcripts signed by trial judges and other checkers have the same legal effect as courtroom transcripts.[48]

Second, the impact of artificial intelligence will redefine all aspects of people's clothing, food, housing and transportation. For example, the development of driverless technology will likely give rise to a smart cab industry, where people in large cities will only need to buy smart travel services instead of keeping their own vehicles. In addition, the application of smart cabs will also have a significant impact on the future of travel difficulties, travel congestion and parking difficulties. In the medical field, the application of artificial intelligence technology will make the current form of medical services change. The future of medical care will develop in the direction of intelligent medicine and precision treatment. With the help of intelligent medicine, daily medical checkups and other services can be done by individuals at home, and only complex medical treatments need to be performed in hospitals, so hospitals will become flattened as institutions providing medical services. In short, the development of intelligent medicine will improve the current problems such as the lack of medical resources and uneven distribution.

Third, the development of artificial intelligence may lead to fundamental changes in the shape of some industries. For example, the insurance industry may undergo significant changes in the future. The development of smart driving technology will dramatically reduce the rate of automobile accidents, thus greatly reducing the significance of automobile insurance. The development of genetic diagnostics and smart

[47] Google's newly developed "Google Neural Machine Translation (GNMT)" technology has dramatically improved its translation standards, making it approach or even surpass human translation standards. See Tencent News: "Machine Learning Pushes Google's Translation Ability Close to Human Level," available at: http://tech.qq.com/a/20161001/022883.htm?t=1475633410576 (in Chinese).

[48] Several Provisions of the Supreme People's Court on Audio and Video Recording of Courtroom Activities, February 22, 2017, available at: http://www.court.gov.cn/fabu-xiangqing-36562.html (in Chinese).

medicine is making people more aware of their lives. This will undoubtedly have an impact on life insurance: on the one hand, customers with disease risks will have their premiums increased or even be denied coverage, thus creating a discrimination problem; on the other hand, people will lack the incentive to buy life insurance when they are fully aware of their health status. In addition, the financial and banking industries will also be affected by the impact of AI.[49] In the banking industry, for example, the application of intelligent robots for bank outlets and the improvement of bank intelligence will make the number of bank branches and bank employees greatly reduced in the future.

Just as artificial intelligence will impact and even reshape certain industries, the employment structure, which is closely related to the industry structure, will also be impacted in all aspects. Many white-collar jobs that were previously considered high-end occupations will be hit harder, resulting in structural unemployment and full-scale unemployment.[50] Therefore, the impact of AI on employment will exert great pressure on social governance in the future. The main groups of unemployment in the future will not only be farmers and workers, but also more highly educated intellectuals and white-collar workers. This is a scenario rarely encountered in social change. On the one hand, peasants are a relatively easy group in the social structure to pacify and accept the status quo, and will not resort to violent methods of social revolution as long as they are given basic social security.[51] On the other hand, the working class has a long history of expressing the pressure of unemployment, and thus society has a long experience in dealing with it. However, human society has very little experience in dealing with intellectual and tertiary unemployment. This will be a great challenge for humanity in the future. Therefore, in these senses, human society is facing a profound revolution with new historical characteristics.

Countries around the world are scrambling to layout artificial intelligence, so artificial intelligence has a profound impact on the world order as well. After the end of the "Cold War", although for a period of time there was an American "unipolar moment", the international pattern more reflects the trend of multipolarity. But in the AI revolution, the multipolar trend after the "Cold War" is likely to be reversed. This is because in the field of artificial intelligence, the United States occupies a super-hegemonic position. Simple arguments can be cited here to prove this point. First, the five largest global companies by market value are all giants in artificial

---

49 "Artificial Intelligence to 'Take Over' Wall Street, Where is the Way Out for Traders?" China Economic Weekly, No. 10, 2017 (in Chinese).

50 Structural unemployment refers to the fact that certain industries will face structural challenges in the short term under the impact of AI, and there is even the possibility of being replaced by historical development. Full-scale unemployment refers to the fact that the coverage of the impact by AI will be all-round. See Qiqi [40].

51 Samuel Huntington, the famous American political scientist, has discussed this profoundly. Huntington pointed out that in the process of great social change, peasants are the most easily pacified group, while intellectuals and students are the most unstable factors in society. See Huntington [41].

intelligence in the United States and have a super voice in the development of artificial intelligence.[52] For example, Google has developed a complete ecosystem in AI, including the developer system TensorFlow, and other giant companies are already competing on the ecosystem, and this competition is internal to the US from an international perspective. Second, the chips needed for AI development are also firmly in the hands of large U.S. companies. Deep learning has greatly enhanced the status of GPUs (graphics processing unit), thus making NVIDIA, the world's most important producer of GPUs, a key actor in the AI era. As the king of chips in the personal computer era, Intel is trying to regain its former super dominance by acquiring the vision processing chip company Movidius. In addition, Qualcomm has also made an important layout in the field of cell phone application AI chips. In short, on the whole, the competition for the top AI chips is still the competition within the United States. Third, in important scenario applications, the United States has an unquestionable first-mover advantage in both primary innovation and value definition. For example, in the driverless scenario, Tesla, Google and others are the most important actors. The U.S. advantage is even more pronounced in the area of smart healthcare. The United States has produced a large number of highly innovative medical innovators in the field of combined gene sequencing and artificial intelligence, such as Illumina, the world's largest producer of gene sequencing instruments.

Compared with the United States, the competitive advantage of traditional Western powers such as the United Kingdom, Germany, Japan and Canada are not obvious. In terms of AI development, the most influential European country is the UK. For example, DeepMind's "AlphaGo" has defeated Lee Sedol and Ke Jie, with its main research and development personnel from the UK. This illustrates the strength of the UK in the field of artificial intelligence in a way. But as DeepMind's acquisition by Google for $400 million shows, it seems that British startups will become subservient to the development of giant American companies. In a sense, this maps out a master–slave relationship between the U.S. and the U.K. in the development of AI. Similar to the UK's position, Israel has a strong presence in the development of AI technologies. The innovative Israeli company Mobileye, for example, was the most innovative company in the field of autonomous driving, but was eventually acquired by the American company Intel.[53] This also reflects the relationship between the United States and Israel. Germany and Japan have a strong industrial base and thus have an advantage in the field of robotics. But since AI is defined by chips and algorithms, the ultimate position in AI development is determined by the platform and ecology. From this point of view, Germany and Japan still have a significant gap in platform and ecology compared to the United States. More special case in the Western countries is Canada. As the most important breakthrough in artificial intelligence in recent years,

---

[52] The five giant companies are Apple, Google, Microsoft, Amazon, and Facebook, collectively known as FAAMG.

[53] On March 13, 2017, Intel Corporation announced on its official website the acquisition of Mobileye, the Israeli company doing the best work in smart driving, for $15.3 billion. available at: https://www.intc.com/investor-relations/investor-education-and-news/investor-news/press-release-details/2017/Intel-to-Acquire-Mobileye/default.aspx.

the three most important experts in deep learning are from Canada.[54] Because of Canada's uniqueness in the layout of AI, giant companies in the United States have set up R&D facilities and industrial layout in Canada. However, although Canada has the original theories and algorithms, it has been unable to develop AI giants due to its small application market, so talent has been flowing to large U.S. companies. In this sense, the development of AI in Canada is still a service to the United States. In short, the gap between the developed Western countries and the United States is widening, not narrowing.

Among emerging countries, only China has benefited from its own large market and research base, and the development of artificial intelligence has seen explosive growth. The gap between the level of AI development in countries such as Russia, Brazil, and India and the United States is likewise widening. It should be especially noted that in the previous the great global division of labor era, developing countries could still gain some development opportunities by participating in the worldwide division of labor. However, in the era of AI, production is mainly produced by robots rather than industrial workers, and such a division of labor may not require the participation of developing countries at all. In addition, with the help of 3D printing technology and AI, the global industrial shift will return to developed countries rather than developing countries. This means that developing countries will share less and less of the global value chain in the era of AI, and even emerging developing countries will face such a dilemma. Therefore, from this perspective, if there is a lack of effective global governance mechanism, those failed countries in the reconstruction of national order may completely lose their development opportunities in front of the technological divide of AI.

In the field of AI, the gap between China and the U.S. is the only one that is closing. Therefore, this section will focus on comparing the advantages that each of China and the U.S. has as a whole. The preceding part of the text has summarized the US's superpower position in the field of artificial intelligence. In this section, we will analyze deeply the advantages of the United States in the field of AI. The author summarizes the advantages of the United States in the field of AI as "three wholes" and "three interactions". The "three wholes" mainly includes the following three points. First, the "whole ecology". The development of artificial intelligence is holistic, and the United States in the field of artificial intelligence advantage is reflected in the layout of the ecology. The so-called ecology, first of all, including the most core artificial intelligence computing chip and other hardware facilities, followed by the chip-based developer operating platform, and finally the application of the developer platform to import a variety of scenarios. And the United States has an unparalleled advantageous position in the ecological aspects. Second, the "whole scene". The United States has a leading position in the application of various scenarios. For example, in the field of driverless, the U.S. has a large number of leading companies and startups such as Google and Tesla. In the field of intelligent medical care, whether it is intelligent detection, intelligent diagnosis or intelligent

[54] The three most important experts in the field of deep learning are Yann LeCun, Geoffrey Hinton and Yoshua Bengio.

treatment and daily medical consultation, the U.S. has many leading technology companies, such as "Watson robot", "Leonardo Da Vinci surgical robot", etc. In addition, in the field of intelligent finance "Kensho" system, the legal field of ROSS system, etc., all reflect the layout and advantageous position of the United States in various application scenarios. Third, the "whole society". It means that the development of artificial intelligence in the United States is not only promoted by enterprises or government, but the result of the interaction of the whole society. In this context, the U.S. AI innovation is not only thinking about the application of a certain scenario, but even thinking about some fundamental and disruptive innovation. For example, since the current knowledge threshold of AI is high and a large number of application scenarios exist in real life, Yoshua Bengio wants to develop a convenient AI system to facilitate people's use.

In addition, the United States has the advantage of interaction mechanism in promoting AI. The main manifestations are as follows. First, the interaction between large and small enterprises. The U.S. has formed a scenario in which giant companies interact well with small companies. For example, giant companies are responsible for building and operating the ecology around chips and systems, while small companies carry out disruptive innovations around a specific scenario. Once the innovation of a small company reaches a large enough social value, the giant companies will acquire small businesses. The huge benefits generated by the acquisition are also an important incentive for small firms to keep innovating. Thus, the interaction mechanism between small and large companies ensures adequate innovation by small enterprises and also allows giant U.S. companies to hold a super voice in the world of AI. Second, the interaction between the civilian and the military. Another important advantage of the United States in science and technology innovation is civilian-military interaction. The primary innovation of major science and technology mainly comes from the military, such as the Internet and computers, which come from the military sector. At the same time, technologies developed by the military sector will be quickly extended to the civilian sector to make them socially beneficial. For example, DARPA has set up competitions in many areas of science and technology to encourage universities, research institutions, and companies to compete. It is in this context that driverless technology has been developed. The military sector's development ensures that its technology is at the forefront, while the social benefits gained by the civilian sector support the updating and iteration of the technology, making it a world leader in both research and development and the transformation of results. Third, the interaction between industry and research: the U.S. R&D sector has strong native R&D capabilities and technology transformation capabilities, and in the field of artificial intelligence to form a vertically integrated model of cooperation between industry and research. This is not only because the giant U.S. companies have an absolute advantage in data, but also they are willing to pay high prices to attract scientists. For example, Facebook has hired Yann LeCun and Google has hired Geoffrey Hinton as chief experts and integrated with top teams. In addition, the flexible education system in the U.S. allows scientists to move freely between academia and industry, which allows scientists to understand both the cutting-edge dynamics of scientific

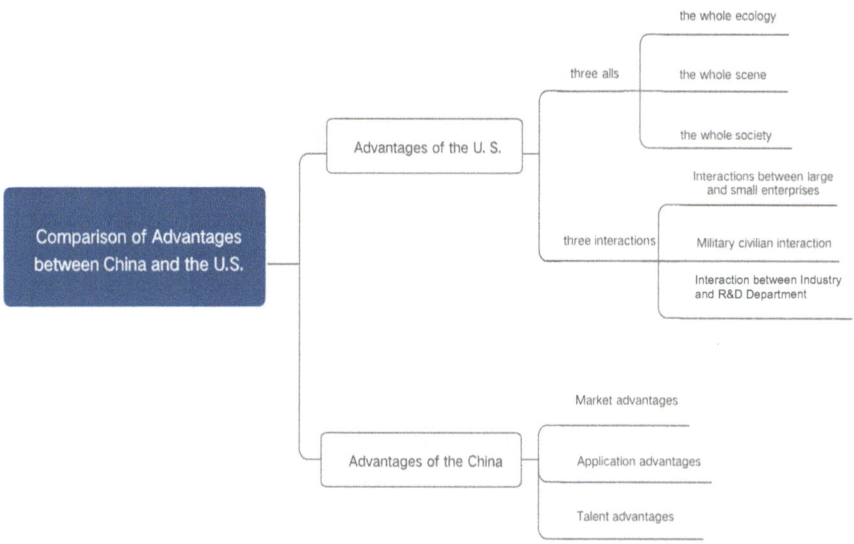

Fig. 1.3 Comparison of advantages between China and the United States in the fourth industrial revolution

research and the actual needs of industry, thus bringing production and R&D closer together.

China's unique advantages over the U.S. are focused on the following points. First, China's large and active market provides a good application environment for technology products. The technical level of Chinese AI products may not be the world leader, but as long as they fit into specific Chinese scenarios, they will have good prospects for development. Second, the basic logic of China is to vigorously promote the application of artificial intelligence and make it serve productivity improvement and economic development. In the process of promoting economic development, Chinese artificial intelligence enterprises can play a certain application advantage. Third, China has a talent advantage. China is now among the world leaders in the total number of international scientific and technical papers published and patents granted. For example, about 1/3 of the scientific research results presented and exchanged at the 26th International Joint Conference on Artificial Intelligence came from China, more than the United States and Europe combined.[55] Chinese science and technology institutions rank relatively high in the global ranking of AI research institutions. For example, in the statistics of the number of papers published by various independent institutions, the Institute of Automation of the Chinese Academy of Sciences made the list as an independent institution with more than 500 published papers, and the citation impact value exceeded the world average, ranking seventh in the world (Fig. 1.3).

[55] Feng [42].

In the area of specific AI technologies, China has a leading position in image recognition and speech recognition. For example, iFLYTEK CO.LTD. is one of the leading companies in the field of intelligent speech. In the field of image recognition, SenseTime, as a leading startup enterprise, raised up to $400 million in B-round financing in 2017, making it the largest single round financing for Chinese artificial intelligence enterprises.[56] This reflects the capital market's preference for artificial intelligence on the one hand, and SenseTime's leading position in visual technology on the other hand. For example, SenseTime has more than one hundred professional talents with PhD and master degree in artificial intelligence, and this talent advantage is also very competitive in the international arena. In addition, AI companies such as MEGVII and CloudWalk are also growing rapidly, which has an important connection with China's active application market. In natural language processing, China also has a number of companies with leading technologies. For example, Xiao-i company invented chatbots much earlier than Apple's intelligent chat tool Siri. In the field of home robots, Ecovacs is a representative Chinese company with the number one share of similar products in Germany.

1.7 Eschatology and the Culture of Superman: America's "Achilles Heel"

In this section, the author will analyze the obstacles to the development of AI in the United States. As mentioned earlier, the United States is currently in a position of absolute dominance in the development of AI, but this dominance is not unshakeable. In other words, there is an "Achilles heel" to this dominant power, i.e., the development of AI in the United States is dominated by "eschatology" and the culture of superman.

First, Western thinking is based on religious culture, which is typical of "eschatological" thinking. Westerners have been brought up with Christian culture, so they cannot break out of this logical cycle. In the West, the disruptive impact of AI on humanity is another demonstration of the end of the world, and therefore they have a pessimistic attitude toward the development of AI. Almost all of the leading American scholars and industrial elites are very pessimistic about the development of AI, such as Elon Musk, Bill Gates, Kevin Kelly, Ray Kurzweil, and others. The source of this pessimism comes from the Christian "eschatology". Because Christianity is diffuse in the United States and the Western world, or rather, all Westerners live in Christianity. Even scientists, who emphasize reason and science, still pay close attention to religious beliefs and life outside of work, and even see scientific progress as another way to prove the greatness of God. Thus, the logic and faith of Western scientists remains a Christian culture. In contrast, the business elite and ordinary

56 Shuang [43].

people are more significantly influenced by Christianity. One of the central narratives in Christian culture is "eschatology", which means that the world is coming to an end due to man's sin and ignorance. Only God will appear at the critical time to save mankind, and the moment of God's appearance is the Messiah.[57] Therefore, it is natural for Christian-influenced scientists and business elites to think about the direction of AI from a pessimistic logic.

According to the logic of Christianity's "original sin", human beings from the bottom of their hearts want to enslave other races or species, and the invention of machines and artificial intelligence is to serve the selfish interests of human beings. At the same time, because of human greed and ignorance, human inventions will eventually prevail over humans as the new masters of the world. James Barrat, for example, predicted that artificial intelligence would be the last invention of mankind.[58] Kurzweil also used "the Singularity is Near" to predict a scenario in which artificial intelligence would overtake humans.[59] This pessimistic sentiment has actually cast a shadow over the development of artificial intelligence in the United States. Musk's view can be seen as ambivalence under the pessimism. Musk is one of the main promoters of AI technology development, but in stark contrast, Musk is also trying to promote AI "eschatology". This typical Western ambivalence will hinder the further development of AI in the United States.

Second, on the basis of "eschatology", the culture of superman has emerged in the development of western AI. This culture of superman is also a projection of Christian culture. The vast majority of American Hollywood blockbusters repeat this type of story. The movie opens with an apocalyptic-like crisis, and the entire plot depicts a Jesus-like superhero saving the world from the crisis. Although these superheroes appear in different images, such as Batman, Captain America, Zorro, etc., their spiritual essence is the same, i.e. a secular version of Jesus Christ. Such superheroes, with Jesus as the object of imitation, have likewise become an important culture in the American business world, as well as a reflection of the culture of Superman. The development of artificial intelligence in the United States is increasingly concentrated in the five giants, such as Google and Apple, and the concentration of superpower has led to what David Korten calls "when corporations rule the world".[60] This has been discussed by Israeli historian Yuval Harari in "Homo Deus: A Brief History of Tomorrow".[61]

---

[57] In Agamben's view, the Messiah is first of all a legal state of being inside and outside; secondly, a special form of time; and finally, the Messiah represents a state of inaction (state of potentiality) in the sense of the Sabbath, while "the Gospel is the vowed form of the Messianic time assumed by the promise". See Agamben [44].

[58] Barratt [45].

[59] Kurzweil [46].

[60] Korten argues that those institutional forces that nurture multinational corporations are the key to the current dilemmas, and that to avoid tragedy, the basic institutions of business must be fundamentally reformed to give power back to small, local businesses. See Cotton [47].

[61] Harari [48].

With this culture of superman dominating, the development of artificial intelligence will undoubtedly intensify the division of society. In recent years, these super giants not only focus on artificial intelligence, but also involved in the field of life sciences. These companies begin to pursue human immortality. In the vision of these giant companies, in the future, humans can "produce" organs through 3D printing technology, and then load human thoughts and emotions into chips and implant them into printed organs, so as to achieve human immortality and re-creation. However, the human immortality pursued by these super giants is not for all people, but a privilege enjoyed by a very small number of super rich people. The rich can enhance the superpower of individuals through technologies such as artificial intelligence, genetic repair, and human enhancement. So under the dominance of this culture of superman, the development of AI will not be a development that considers all people, but more of a game for the technological and capital elites.

The pessimistic attitude of Western Christian "eschatology" brings the development of artificial intelligence into a paradoxical state. Imagine if the development of AI is the self-destruction of human beings, is there any need for development? In addition, the culture of superman does not consider the development of AI from the perspective of society as a whole. Technological supermen look at economic benefits and disruptive effects, but lack consideration of the wave of unemployment caused by AI development. If the negative effects of this AI development are not limited and mediated, it will trigger a situation where angry unemployed people smash the machines, which will eventually lead to social unrest and civilizational regression.

Based on the obstacles to the development of AI in the United States, this chapter concludes with an analysis of the possibility of a breakthrough in China in the future. The author believes that the greatest advantage of China's future development of AI is culture. Completely different from Western culture, Chinese culture has two characteristics that can mediate exactly the dilemma of Western culture.

First, symbiosis theory. Unlike "eschatology", symbiosis theory emphasizes the symbiosis between human beings and the world, and among species. Chinese culture places great emphasis on the harmony between human beings and other living beings, the most representative of which is the Buddhist saying, "While sweeping the floor, you should be very careful lest any ants be killed. While lighting a candle, you should be cautious lest any moths be burned to death." The main logic of Western thinking about artificial intelligence is Hegel's master–slave dialectic, that is, the machine is first a slave of human beings, and as the machine can not stand human slavery, it eventually becomes the master by fighting to overcome human beings. Unlike the Western master–slave logic, the logic of Chinese culture believes that even if machines are human creations, human should respect them as much as they respect ant's life, thus achieving a state of harmonious symbiosis between human and machines. Humans need the help of machines in solving problems, while machines achieve subject status and value in the process of helping people. Thus, the future of artificial intelligence is no longer the end of the world, but the long-term harmonious coexistence of human and machine.

Second, humanistic culture. China considers issues from the perspective of the people as a whole, rather than from the perspective of elites. In sharp contrast to the Western culture where the fate of the world is in the hands of supermen, Chinese culture emphasizes the value of each individual. For example, Mencius, the representative figure of Confucianism, believed that "all man can be sages like Yao and Shun".[62] Chairmen Mao Zedong, in his praise of the working people, also noted that "China with a population of 600 million has lots of heroes".[63] In fact, Chinese leaders think from the standpoint of the majority rather than the elite minority, as President Xi Jinping pointed out, "The people's desire for a better life is our goal".[64] The "targeted poverty reduction" and "building a moderately prosperous society in all respects" proposed in recent years are manifestations of this thinking. From China's perspective, artificial intelligence is not for the profit of a few, but for the improvement of the overall welfare of the vast majority of people. Thus, when developing AI, China will consider various development risks and take targeted measures to address them. For example, when the development of AI leads to mass unemployment in some industries, China will adopt social insurance, AI taxes and transfer payments to provide subsidies and relief. In fact, only a view of AI development that takes into account the overall interests of all people is harmonious and has the lowest social cost.

Otherwise, another advantage of China's AI development is its special position in world politics, i.e., China is in the middle ground between developed and developing countries. On the one hand, China is the hope of AI among developing countries. Because AI is developing so fast that developing countries cannot get on the AI development fast train. And with the end of localization and global integration of production brought by AI, it will be more difficult for developing countries to catch up with developed Western countries in the future. On the other hand, the gap between China's AI development level and that of developed countries is not out of reach, while the technological may divide the world apart. As a representative of developing countries, China is able to assert the interests and demands of developing countries in the AI era.

Therefore, China's development of AI has a moral stance, i.e., China develops AI not only for its own sake, but also for the development rights and interests of developing countries at large. In the era of AI, the gap and technological divide between developed and developing countries is widening rather than narrowing. The overall gap between developing countries and the United States in AI-related technology, culture, and facilities will continue to widen, and even the possibility of development opportunities through participation in the world division of labor system will disappear. China, on the other hand, has the potential to be the hope for developing AI in developing countries. In fact, China is also actively fighting for the interests of developing countries in world politics. For example, China's promotion of the "Belt and Road Initiative" and Asian Infrastructure Investment Bank is to share its experience in infrastructure development and national governance with developing countries, and to share the fruits of China's reform and opening

[62] Mencius—Gaozi(ii).

[63] Mao Zedong, Getting Rid of the God of Plague (in Chinese).

[64] Jinping [49].

up through joint construction. The concept of community with a shared future for mankind is a sublimation and summary of this holistic view. When China thinks about international political issues, it does not only consider China's own interests, but also the interests of the vast number of developing countries. In this sense, therefore, China's development of artificial intelligence technology has a special mission in world politics.

The current development of artificial intelligence worldwide has seen a unipolar pattern in which the United States is the only country, and even the traditional Western powers have a large gap compared to the United States. The super power of the United States is mainly reflected in its control and monopoly of core technologies such as chips, and the overall construction of the ecology such as operating systems. In a long time, the United States in the field of artificial intelligence super position will be difficult to shake. But at the same time, there is also a fundamental weakness in the superpower of the United States, mainly reflected in the Christian culture and its projection of pessimism into the development of artificial intelligence. So American AI development is characterized by two things: a self-contradictory pessimism; and a superman culture that does not take into account the overall development. These two characteristics actually imply the fate of the United States in the future development of artificial intelligence end. So in the short and medium term, the U.S. super position in AI seems unshakable, but in the long term, the foundation of the culture of AI established by U.S. is unreliable.

In contrast, the ideological characteristics of Chinese culture can precisely compensate for the two tendencies that currently pervade the development of AI. On the one hand, the concept of "Man is an integral part of nature" can help people build a harmonious relationship between human and AI, and at the same time, it can make people look at the development of AI with a relatively optimistic mindset. Because of this attitude, China will become a positive driving force and application market for AI development. On the other hand, a view of AI development from traditional Chinese culture is the representation for interests of the greatest number of people rather than a few elites. This view of development will both actively promote the development of AI and comprehensively examine the disruptive impact and risks that AI brings to human development. In the era of AI, there is a need for synergy rather than mutual tension between people and AI, and between those in dominant and vulnerable positions.

In fact, the view of AI development from traditional Chinese culture answers three major questions of AI development: first, the question of whether to develop AI; second, the question of for whom to develop; and third, the question of how to develop AI. In the face of these three major questions, the answers given by Western culture are confusing. On the question of whether to develop AI, according to the Western Christian culture, the idea is not to develop AI, while in fact the industry is actively promoting it. This actually creates a paradoxical situation. On the question of who to develop for, Western culture points to the result of ultimately developing for an elite few. As for how to develop AI, Western industry is still trapped in confusion and dilemma. But the answer to the question of AI development from traditional Chinese culture is clear. Firstly, we need to develop artificial intelligence, as it can coexist harmoniously with humans under appropriate constraints. Secondly, we need to develop artificial intelligence for the entire people rather than for a few elites, and from the perspective of a community with a shared future for mankind, this even has global significance. China has a very strong moral position because it is the hope for developing AI in developing countries. In confronting the third issue, Chinese culture reminds humanity to always pay attention to and guard against the unemployment and ethical problems brought about by AI. Only preventive development can effectively curb the occurrence of systemic risks, so that AI can serve to enhance human welfare and build a community with a shared future for mankind.

References

1. Marx K (1998) Frederick Engels: collected works, vol 31. People's Publishing House, Beijing, p 94 (in Chinese)
2. Qixuan H (2013) Power shift and technological change in great powers. Shanghai Jiao Tong University Press, Shanghai (in Chinese)
3. Leften Stavros Stavrianos (2015) A global history. Peking University Press, Beijing (in Chinese)
4. Diamond J (2000) Guns, germs, and steel: the fate of human societies. Shanghai Translation Publishing House, Shanghai, p 262 (in Chinese)
5. Yucang W (2004) History of science and technology. China Renmin University Press, Beijing, p 316 (in Chinese)
6. Zhenhuan J (2017) General history of technology. China Social Sciences Publishing Houses, Beijing, pp 213–214 (in Chinese)
7. Basara G (2000) The evolution of technology. Fudan University Press, Shanghai, pp 50–53 (in Chinese)
8. Clapham JH (1964) History of modern economy of the United Kingdom (Volume I-II). The Commercial Press, Beijing, pp 604–609 (in Chinese)
9. Britain G (1807–1869) The statutes of the United Kingdom of Great Britain and Ireland. His Majesty's Statute and Law Printers, London, p 101
10. Lay M (1999) Ways of the world: a history of the world's roads and of the vehicles that used them. Rutgers University Press, New Jersey, pp 138–141
11. Jianhong D (2002) General history of Germany. Shanghai Academy of Social Sciences Press, Shanghai, pp 330–335 (in Chinese)
12. Dongliang Y (2010) Modern economic history of Japan. World Affairs Press, Beijing, pp 72–107 (in Chinese)
13. Hongsheng S (2012) Hidden participants in the Japanese invasion of China - Japanese enterprises. Res Hist Jpn's Invas China 1 (in Chinese)
14. Jianhua H, Yiyun T (2015) Frontiers of international smart phone technology and its implications for China's technology strategy. Sci Technol Manag Res 11 (in Chinese)
15. Maoxin L (2011) Trend analysis of cross border migration of professional and technical talents after World War II. J Hist Sci 12 (in Chinese)
16. Goodman S (1979) Soviet computing and technology transfer: an overview. World Polit 31(4):540
17. Weiying Z, Bin S (2004) On entrepreneurs - the kings of economic growth. Sdx Joint Publishing Company, Beijing, pp 16–17 (in Chinese)
18. Adirim I (1991) Current development and dissemination of computer technology in the Soviet economy. Sov Stud 43(4):651–667
19. Fei Y (2017) Industrial development analysis of Japanese semiconductors. Appl Integr Circuits 1 (in Chinese)

20. Woo-Cumings M (2008) The developmental state. Jilin Publishing Group, Ltd., Changchun, pp 5–13 (in Chinese)
21. Junwen F (1993) Analysis of Japanese foreign capital flows and China's use of Japanese capital in the 1990s. Soc Sci 4 (in Chinese)
22. Yimin G (2002) The neoliberal side of educational reform in Japan. Tsinghua Univ Educ Res 6 (in Chinese)
23. Ohmae K (2018) Low desire society: the new - wealth of nations in the age of loss of great ambition. Shanghai Translation Publishing House, Shanghai, pp 38–50 (in Chinese)
24. Wenqing Y, Rundong L (2016) Innovation and reference of Japanese pension model in the context of aging. Zhejiang J 1, (in Chinese)
25. Marx K (1972) Frederick Engels: complete works, vol 2. People's Publishing House, Beijing, p 184 (in Chinese)
26. Lenin collected works, vol 27. People's Publishing House, Beijing, p 401 (in Chinese)
27. Dos-Santos T (1999) Imperialism and dependence. Social Science Literature Press, Beijing, p 302 (in Chinese)
28. Baer W (1972) Import substitution and industrialization in Latin America: experiences and interpretations. Lat Am Res Rev 7(1):95–122
29. Huntington S (2008) Political order in changing societies. Shanghai People's Publishing House, Shanghai, pp 160–219 (in Chinese)
30. Lin JY, Fang C, Zhou L (1999) Comparative advantage and development strategy: re interpretation of the 'East Asian Miracle'. China Soc Sci 5 (in Chinese)
31. Nenni D, Dingee D, Unleashed M (2015) The origin and evolution of ARM processors in our devices. CreateSpace Independent Publishing Platform, South Carolina, pp 131–150
32. Jun W (2016) International practice and theoretical reflection on industrial policy transformation. Econ Res Ref 28 (in Chinese)
33. Lipset SM, Lenz GS (2001) Corruption, culture, and markets. In: Harrison L, Huntington S (eds) Culture matters: how values shape human progress. Basic Book, New York, pp 112–124
34. Schwab K (2016) The fourth industrial revolution. CITIC Press Group, Beijing, pp 11–15 (in Chinese)
35. Erisman P (2015) Alibaba's world. CITIC Press Group, Beijing, pp 72–77 (in Chinese)
36. Xiaobo W (2017) Tencent biography. Zhejiang University Press, Hangzhou, pp 35–44; 66–71 (in Chinese)
37. Jiangan L et al (2016) Research on innovation paths of business model under 'internet +' formats based on ground theory—a case study of Di Di Taxi. Soft Sci 7, (in Chinese)
38. Santos T (1970) The structure of dependence. Am Econ Rev 60(2):231–236
39. Qiqi G (2018) Artificial intelligence: taming the Severtan. Shanghai JiaoTong University Press, Shanghai, pp 119–124 (in Chinese)
40. Qiqi G (2017) Artificial intelligence is not just a 'buzz' belonging to the tech world. Liberation Daily, June 6, (in Chinese)
41. Huntington S (1996) Political order in changing societies (Trans by Guanhua W, Wei L et al). SDX Joint Publishing Company, Beijing, pp 254–273 (in Chinese)
42. Feng L (2017) World AI stage shines with Chinese landscape. People's Daily, August 26, (in Chinese)
43. Shuang CH (2017) SenseTime completes $410 million series B financing, setting a record for a single round of financing in AI. Xinjing News, July 11. http://www.bjnews.com.cn/invest/2017/07/11/450047.html (in Chinese)
44. Agamben G (2011) The remaining time - interpreting the book of romans. (Trans. by Liqing Q). Jilin Publishing Group, Changchun, p 114 (in Chinese)
45. Barratt J (2016) Our last invention: artificial intelligence and the end of the human era (Trans. by Jia L). Electronic Industry Press, Beijing (in Chinese)
46. Kurzweil R (2011) Singularity is near (Trans. by Qingcheng L, Zhenhua D, Yuan T). Machinery Industry Press, Beijing, pp 12–14 (in Chinese)
47. Cotton D (2006) When corporations rule the world (Trans. by Daoyong W). Guangdong People's Publishing House, Guangzhou, pp 11–12 (in Chinese)

48. Harari Y (2017) A brief history of the future: from homo sapiens to divine man (Trans. by Junhong L). CITIC Publishing Group, Beijing, pp 277–317 (in Chinese)
49. Jinping X (2012) People's aspiration for a better life is our goal. People's Daily, November 15 (in Chinese)

Chapter 2
Intelligent Risk Governance and Intelligent Society Building

The development of artificial intelligence is closely related to the Internet. Artificial intelligence has entered the third wave of overall growth, and the rise of this wave is closely associated with the following factors. First, due to the development of Internet technology, the amount of data formed on the Internet is increasing, providing a database for AI development. Second, the computility of various intelligent devices is rising. Third, algorithmic models, profound learning algorithms, have made significant breakthroughs in recent years. In terms of these three points, the first two closely reflect the essential supporting role of Internet development for AI. In terms of the future, the further development of AI also needs to lend a hand with the Internet. The story of the mobile Internet has made the smartphone the most essential tool. The ever-increasing performance of cell phones and the rapidly increasing amount of mobile data also boost AI. These two also explain the coherence and continuity between China's "Internet + " strategy and the current AI development strategy. In addition, implementing 5G, a new generation of communication standards, will also significantly contribute to the further development of AI. For example, driverless-ness is a crucial part of the application scenarios of AI. Since the three main features of 5G include ultra-low latency, this makes the intelligent driving platform more conducive to handling emergencies during driving.

2.1 Taming the "Seviathan" of Artificial Intelligence

What is artificial intelligence? Let's try to give it a definition. Artificial intelligence imitates human intelligence and seeks to achieve specific tasks. Artificial intelligence consists of three aspects: the first is computational intelligence, which involves fast computation and memory storage capabilities. In the view of computer scientists, artificial intelligence is first and foremost computational behavior, i.e., it involves data, arithmetic power, and algorithms; the second is perceptual intelligence, which

Q. Gao and J. Zhang, *Artificial Intelligence Governance and the Blockchain Revolution*, Artificial Intelligence and the Rule of Law,
https://doi.org/10.1007/978-981-99-9211-9_2

consists of the machine's perceptual capabilities such as vision, hearing, and touch, i.e., the device can collect and analyse information about the environment through various types of sensors and, after processing, responding and reacting rationally on demand. The third is cognitive intelligence, which refers to the ability of the machine to think independently and solve problems. Nowadays, AI mainly stays at the first and second levels, and cognitive intelligence is complicated to achieve because it involves deep semantic understanding.

In my talks, I often ask the audience when they probably learned about AI. Many people will talk about the duel between AlphaGo and Lee Sedol which is what most people in China understand AI to be. Most Chinese people understand the first wave of AI as the third wave of AI in the world. From a worldwide perspective, artificial intelligence has developed in three waves. The first wave was mainly between 1950 and 1970, when the main work was done by computer scientists working on machine reasoning systems, and the theoretical school of thought in this period was called symbolism. Of course, neural networks and expert systems were invented in the earliest part of this period. However, there was a great desire in the United States to make some practical applications of AI, such as machine translation. Still, the technology at that time could not do it, so this wave stopped in disappointment. The second wave was from 1980 to 2000. In the second wave, concepts such as the statistical school, machine learning, and neural networks, which we are currently talking about, were introduced. The dominant theoretical school in this wave was called connectionism. The third wave came after 2006, thanks to the promotion of big data, in which Google played an important role. Google's successful prediction of influenza with big data, which caught the attention of the US health authorities, is the most important example of how big data and artificial intelligence are closely linked.[1] During this wave, AI technology and its applications have improved tremendously. The computility has increased, the amount of data has increased, and there have been considerable breakthroughs in algorithms centered on neural networks, so AI has entered people's lives under the aegis of big data. That is, we Chinese feel is in the first wave of artificial intelligence, but from a worldwide perspective is currently in the middle of the third wave.

Three related concepts about AI need to be clarified: the first is ANI (Artificial Narrow Intelligence), the second is AGI (Artificial General Intelligence), and the third is ASI (Artificial Super intelligence). ANI is specialized AI. It isn't easy to use it directly in other scenarios. What is the ideal of many scientists now? It is to do AGI, that is, general AI so that the AI can have more applications and be migrated into other scenarios. So, what is ASI? As the name implies, it is intelligence that exceeds human intelligence. ASI doesn't exist yet, and we hope it never will, or it will be subversive to the meaning of humanity.

Many experts have discussed the possible future of AGI. For example, Ray Kurzweil, an American scientist and inventor, believes that AGI could surpass humans by the 30 s or 40 s of this century and sees this day as Singularity. Kurzweil also set up the highly innovative Singularity University for this purpose, to promote

[1] Ginsberg, Mohebbi et al. [1], pp. 1012–1014.

rapid human progress to avoid AI surpassing humans. Similar views about Kurzweil in the West are not just a few. For example, Nick Bostrom, a leading futurist at Oxford University in the UK, believes that ASI can overtake humans.[2] Bostrom argues that ASI comes in many forms and that such intelligence could acquire universal intelligence and become a full-fledged threat to or replacement for humans. Bostrom divides ASI into three forms: speed superintelligence, collective superintelligence, and quality superintelligence.[3] Speed superintelligence is similar to the human brain but faster than the human brain intelligence. In Bostrom's words, speed superintelligence is intelligence that can do everything human intelligence does much quicker. Bostrom believes that such a whole-brain simulation system could read a book in seconds if it were 10,000 times faster than the human brain. If such a system were 1 million times faster than the human brain, a ASI could do 1000 years of human intellectual work in one working day. Collective superintelligence is composed of a vast number of tiny intelligence, and in many general-purpose areas, the overall performance of such intelligence dramatically exceeds that of existing cognitive systems. Collective superintelligence is best at solving problems divided into subproblems and can find and verify solutions to each subproblem individually and simultaneously. According to Bostrom, collective superintelligence can be integrated in a way that is both loose and tight, i.e., it can form a unified intelligence. Quality superintelligence is a system that as fast as the human brain but with a qualitatively superior level of intelligence compared to humans. According to Bostrom, the gap between this quality superintelligenc and human intelligence is like that between human intelligence and the intelligence of elephants, dolphins, and orangutans. Bostrom believes that future superintelligence could acquire a new set of cognitive modules and enable general intelligence to gain unique advantages by constructing complex knowledge engineering.

Why is artificial intelligence so important? Because what AI represents is the fourth industrial revolution. In this industrial revolution, the most critical technology is AI.[4] Of course, some related technologies exist, such as the Internet of Things, blockchain, supercomputing, brain science, etc. But because of the significant role of AI in it, this revolution is called the Smart Revolution. This means that AI is not a simple technology but a strategic technology, a key force in this technological revolution and industrial change round. Artificial intelligence's good or bad development determines whether we have the initiative in the competition of this industrial revolution. Artificial intelligence has already significantly impacted all aspects of social life. For example, the impact of AI on transportation lies in making driverless-ness a possibility. The future of transportation will reflect three elements more: the first is new energy, the second is driverless, and the third is shared mobility. Based on AI technology, people don't need to own a car. Owning a car means getting from one

---

[2] Moreover, Bostrom pointed out that "Artificial intelligence already outperforms human intelligence in many domains." See Nick Bostrom, *Superintelligence: Paths, Dangers, Strategies*, Oxford: Oxford University Press, 2014, p.11.

[3] Nick Bostrom [2], p. 52.

[4] Velarde [3], pp. 41–48.

place to another smoothly, conveniently, and economically. Still, thanks to driverless technology, people may no longer need to own a car, instead, they can buy travel services directly from the place of departure to the area of destination. Today's automakers may need to shift to the role of driverless service operators in the future. In this process, automakers must fully integrate and collaborate with today's ride-hailing software and HD mapping companies. Shared mobility also enables many of our previous structural problems to be solved. For example, parking has always been one of the most challenging problems in urban governance, and this problem may be strongly solved in the context of shared mobility. Because cars are traveling on the road without having to park somewhere, this could free up many parking resources to be used as greenery or places for people to rest.

Healthcare is also a problem that plagues humanity. The biggest problem with healthcare is that there are never enough medical resources for the demand, and artificial intelligence can significantly expand this resource. The advantage of AI is that it can gradually automate some of the tasks established doctors have traditionally done. For example, AI can help doctors in imaging departments do a better job.[5] The thinking here must be that AI is an aid to humans, not a replacement. Because medical resources will always be scarce for humans, many doctors spend much time doing low-end, repetitive work. Therefore, when AI does this repetitive work, doctors can do better scientific research, communicate with patients more gently, understand the condition in depth, make better treatment plans according to the exact condition of each patient, etc. In addition, AI can also expand medical resources to poorer areas. As soon as AI technology achieves a breakthrough, its problem solution is that it will be relatively stable. In addition, telemedicine becomes possible based on 5G technology, then seriously ill patients can be operated on or treated in remote areas through remote devices. In addition, artificial intelligence technology can be used for drug development, which can help people to develop new drugs to treat complex diseases at a lower cost and in less time.

The tremendous significance of artificial intelligence for education is its ability to promote an adaptive approach to education. Our current education is still multi-personal because the teacher's time is limited, so teaching each child according to their personality and characteristics is challenging. However, the artificial intelligence system can better understand each child's educational status. It can significantly improve learning efficiency by continuously understanding each child's condition more precisely through an adaptive system. In addition, AI can also play a massive role in teacher support and academic assessment.[6]

On a worldwide scale, the development of AI may have important implications for the future structure of the world. On the one hand, the U.S. superpower in AI remains strong. For example, the U.S. has a decisive advantage in general-purpose computing hardware. NVIDIA's GPUs, Google's TPUs, and Qualcomm's smartphone chips have clear benefits. In addition, U.S. companies and universities have a decisive advantage in the algorithmic framework of artificial intelligence.

---

[5] Hosny, Parmar, Quackenbush et al. [4], pp. 500–510.

[6] Goksel and Bozkurt [5], pp. 224–236.

Among some specific application scenarios, for example, in special robots, Boston Dynamics’ technological advantage is also apparent. In autonomous driving, Google and Tesla are also leading companies. In addition, the United States is also relatively obvious in some of the original innovation field advantages.

China has been developing faster in recent years in artificial intelligence. We have kept pace with developed countries in the layout of 5G communication and, in some aspects, are even ahead of them. At the same time, under the active guidance of the national grand strategy, for example, after the State Council issued the “New Generation Artificial Intelligence Development Plan” in 2017, all ministries and commissions have also issued relevant AI promotion plans, and provinces and cities have also formulated appropriate promotion programs. China’s most significant advantage is its vast market and relatively regular consumer base. China has a large population and a high degree of information technology, which are the basis for the rapid application and implementation of AI in various scenarios in the future. Overall, China still has some shortcomings in bare chips, algorithm framework, and ecology, and there is much room for improvement in basic theory and native innovation. In the new technological revolution, other developing countries may need clarification and clarification. On the one hand, developing countries can use new learning methods and open-source software to cultivate relevant talents and significantly shorten the time to catch up with developed countries. At the same time, intelligent approaches can also help developing countries overcome the limitations of traditional culture. But at the same time, the intellectual revolution may further widen the gap between developing and developed countries.

The critical problem now is that developed Western countries are not willing to transfer cutting-edge technologies such as artificial intelligence to developing countries, while when developing countries achieve certain breakthroughs in these emerging technologies, developed countries will also use various methods, including investment reviews, export controls, and restrictions on science and technology and personnel exchanges to hinder the scientific and technological progress of emerging countries. Developed countries hope to keep these technologies in their own hands for a long time and prevent the spread of technology to dominate a new round of scientific and technological revolution. In this sense, developing countries, especially those with weak foundations, will be stuck in a dilemma. Those developing countries that use labor as a competitive advantage will be further in a more marginal position. As developed countries will pull back their industries and drive production through machines, these developing countries with weaker bases will have less and less opportunity to participate in international competition and world production.

Developed Western countries mainly define the world’s legal rules, policies, and principles related to AI. For example, the most influential “Asilomar’s 23 Principles of Artificial Intelligence” was formed by Western entrepreneurs such as Elon Reeve Musk. In addition, the most influential Isaac Asimov’s “Three Laws of Robotics” in AI was first proposed by an American science fiction writer and has become the essential principle in machine ethics in the West. In June this year, China released the “New Generation of AI Governance Principles.” This is the first time a developing country has proposed AI-related governance guidelines, so the release of this guideline is of great significance. From an objective point of view, the number of international rules on AI that developing countries have participated in formulating is still relatively small in the international community. With the further development of China in AI and other fields, then similar rule-making will become very important.

2.2 Risk Governance and Regulation of Artificial Intelligence

An essential feature of this round of AI is data-driven intelligence. Therefore, three important types of risks may arise during this round of AI, as follows.

First, technology risks. Artificial intelligence technology itself has some risks. First, there is the problem of algorithmic black box. At present, artificial intelligence technology is mainly reflected in the algorithm of deep learning technology deepening, and the technology of deep learning itself has the problem of the algorithmic black box.[7] Because of the un-interpretability in the neural network, it is difficult to use this technology in some irreversible decision-making areas. For example, in areas such as justice and healthcare, once the final decision is made through an algorithm, that un-interpretability becomes a severe problem. Because intelligence cannot make adequate and practical explanations, medical diagnoses or judicial decisions based on such are hardly convincing.

Simultaneously, technology is subject to cyclical risk. There will be development cycle in any development of technology, the industry is also the same. The current development of artificial intelligence, from a worldwide perspective, has three ups and two downs; that is, there are two trough periods. And in this development of artificial intelligence, it isn't easy to judge when the trough period will appear.

There is also the problem of the cost–benefit ratio of the technology. Many AI products are currently very effective, but they cannot be mass-produced or mass promoted because the price of the products needs to be lowered. For example, Boston Dynamics' humanoid robot technology is very advanced. However, the mechanical skeleton joints of Boston Dynamics robots need to be realized through 3D printing, so mass production still needs to be improved. Another example is AlphaGo, which plays chess with Lee Sedol and Ke Jie and costs thousands of dollars per game. Therefore, in practical applications, these technologies will face enormous cost challenges.

Second, privacy risk. Because current AI development is highly dependent on big data, today's AI companies are, first and foremost, data companies. As a result, the leading AI companies are racking their brains on how to collect data on various dimensions. Many investment companies are also basing their judgments about the future of a company or project on the amount of data available to the product. Through

[7] Carabantes [6], pp. 309–317.

some overbearing design, some companies' excessive collection of personal information can lead to serious privacy breaches.[8] In December 2017, Consumer Watchdog, a U.S. consumer protection organization, issued a report. The organization's research claimed that Amazon and Google's smart speakers might be used to listen to large amounts of information about users and use it for advertising promotions.

In addition, some foreign studies have suggested that China will make high-speed progress in big data and artificial intelligence because it does not pay attention to privacy protection. This is a wrong judgment and may also mislead China's AI development. This is because the public, inspired and induced by certain events, may resist many technological advances. The gene editing incident, for example, reflects such a characteristic. Therefore, AI companies should take this issue seriously and always appreciate the Chinese public's importance of privacy rights. The government should also actively introduce appropriate legal regulations, which could prevent some public opinion incidents and thus protect technology companies in general. Privacy issues are more critical to explore in the field of innovative healthcare. Because of the unique nature of medical data, it is inherently controversial whether data rights should belong to individuals or companies. The General Data Protection Regulation introduced by the EU tries to define this issue effectively, but the relevant laws in China still need to be made clear.

Third, security risks. Artificial intelligence technology itself can be used in legitimate areas. For example, in recent years, at the concert of Jacky Cheung, China's public security authorities arrested many suspects through facial recognition technology. In addition, artificial intelligence technology can also be more effective in identifying whether the picture has been forged or changed, etc. But at the same time, AI technology can also be used for forgery. For example, voice synthesis technology can be used to forge voices. Criminals can develop a person's voice through technology by having the person's voice information. These voices may be used for fraud, blackmail, and other purposes. In addition, criminals can also use techniques that generate fake videos. For example, a faked video of Obama attacking Trump was circulated online. The release of this video quickly caused a significant stir in the United States and other countries. Thus, AI technology has provided criminals with new ways to commit crimes while preventing them.[9] The motivation for AI to be used for crime is more apparent due to several issues, such as the black economy, so we cannot underestimate the security risks arising from AI (Fig. 2.1).

As human society gradually enters the intelligent community and many innovative products enter the home, security risks will become more prominent. Previously existing cyber security issues will be more significant under the fashionable society. Once, some cyber hackers manipulated other people's computers remotely through malicious programs to assist themselves in Bitcoin mining. However, when many self-driving cars emerge, these hackers are more likely to break into driverless systems, creating a considerable risk. This scenario has been seen in movies and TV shows such as "Robo-Ki" and "Terminator." In addition, more sophisticated

[8] Shank and Gott [7], pp. 1636–1645.

[9] Caldwell, Andrews, Tanay and Griffin [8].

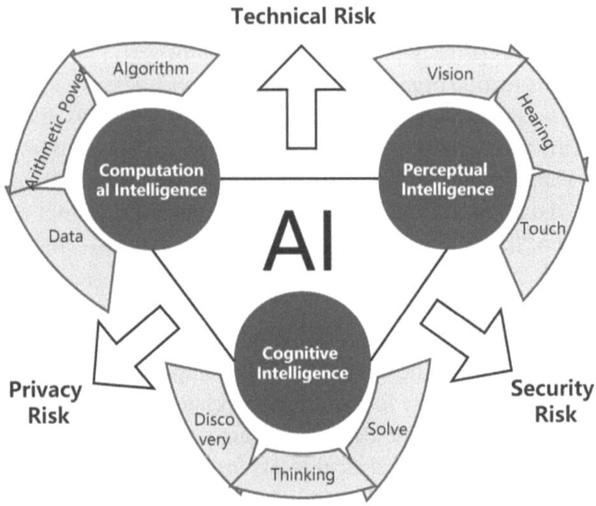

Fig. 2.1 Composition and risks of artificial intelligence

and intelligent medical devices will be used more widely in human treatment. For example, nanobots will enter the human blood vessels and treat certain diseased areas in the human body more precisely. But likewise, hackers can quickly kill people by tapping into these systems. In the age of intelligence, therefore, the social risks arising from cyber-hacking may become even more significant.

The author believes the focus should be on two aspects to regulate the social risks arising from AI.

First, shifting from increasingly popular computational departmentalism to humanism is essential in philosophy. Computational departmentalism can also be called computationalism, which means that computation is the center. There are three elements of computing, computility, algorithm, and computing material. The computility is the computing power, the algorithm is the method and model to solve the problem, and the computing material is the data collected by the sensors we are discussing today. As human society moves more and more towards an intelligent society, people's reliance on data has become more apparent. Algorithms help people make various decisions through data analysis. People will reduce the cost of computing by continuously improving computility, while people's behavioral trajectories will be recorded as computing material. Therefore, a tight cycle will be formed between these three, making people pay more and more attention to computation. However, in this process, it is easy to forget the purpose of computing. This is what the author emphasizes as the meaning of humanism.

There is already a tendency to be algorithm-centric. In many areas, as intelligent devices enter people's daily lives, people rely heavily on algorithms to make decisions. For example, the advent of navigation systems has made it possible for young people today to learn how to navigate and remember directions. Therefore, the emergence of intelligent devices will, to some extent, make many human capabilities in the degradation. In the long run, some of the basic survival capabilities of humans may be lost entirely. This overall trend will bring terrible results for human beings. Therefore, when we use artificial intelligence, we still have to interrogate constantly: why do we use these intelligent systems? What is the purpose of using these intelligent systems? What problems can intelligent systems help us solve, and what problems can they not solve? The most crucial issue is decision-making. If we leave many decisions to algorithms, we may lose the centrality of human beings.

Second, at the technical level, the development of blockchain is imperative. Developing blockchain technology is the most effective way to solve the two significant privacy and security risks. The author believes that artificial intelligence and blockchain are two sides of the same coin of intelligent technology. Artificial intelligence achieves the intelligence of things, while blockchain achieves the intelligence of relationships. The continuous development of artificial intelligence will help the breakthrough of blockchain technology. At the same time, blockchain can circumvent many of the problems faced by the development of artificial intelligence. For example, the issue of privacy protection can be effectively solved by asymmetric key technology. In addition, the traceability and immutability of blockchain will help solve the cyber security problem.[10] Many scientists and observers see blockchain as the future direction of Internet development. In other words, blockchain is the core of the new generation of Internet technology.

China is increasingly at the forefront of the world, both on the Internet and in artificial intelligence. Especially in this round of artificial intelligence development process, many artificial intelligence applications are likely to be the first to land in China. The development of intelligent security is the most typical example. In addition, China is expected to excel in innovative healthcare and brilliant driving areas. Thus, the development of AI technology has led China to enter "no man's land" in some technology regions, making risk governance and related rulemaking critical. However, the absence of mature risk prevention and control systems or measures worldwide puts enormous pressure on China to innovate in the Internet and AI. Therefore the role of social science in this becomes more significant. In previous articles, the author discussed intelligent social science and the new liberal arts. To be fully developed, artificial intelligence needs not only the disciplinary construction and efforts of intelligent science and engineering but also the escort of intelligent social science. In the ninth collective study of the Political Bureau of the CPC Central Committee in 2018, President Xi Jinping emphasized the need to integrate multidisciplinary forces, strengthen research on legal, ethical, and social issues related to artificial intelligence, and establish sound laws, regulations, institutional systems, and ethics to guarantee the healthy development of artificial intelligence. Therefore, it is the right time to build a new liberal art in the new era. Social sciences should not only answer traditional classical propositions but also play an important early warning and guarantee function in developing and landing emerging technologies (Fig. 2.2).

[10] Taylor, Dargahi, Dehghantanha et al. [9], pp. 147–156.

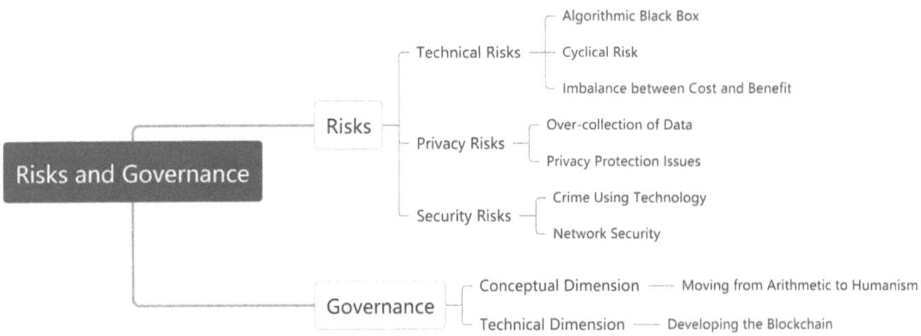

Fig. 2.2 Risks and governance of artificial intelligence

2.3 Legal and Ethical Developments in Automated Driving Technology

Autonomous driving has made significant technological breakthroughs in recent years. Many independent driving development companies such as Baidu, Google, and Tesla have taken 2020 as an important year for autonomous driving to take hold. As 2020 approaches, we will find that the current problems facing autonomous driving include technical difficulties and many legal and ethical issues.

Regarding the legal issues of autonomous driving, the first is the issue of self-driving cars on the road, which involves road traffic safety laws. According to the current Chinese road traffic law, self-driving cars cannot be driven on the road without approval. At present, only some ministries and local governments have introduced some policy regulations; for example, Beijing has introduced the first domestic self-driving regulations, "Guidance on Accelerating the Work Related to Road Testing of Self-driving Vehicles" and "Implementation Rules for Road Testing Management of Self-driving Vehicles." These two guidance documents conditionally allow testing of self-driving cars on specific roads. Autonomous driving is divided into L0 to L5, with six levels corresponding to no autonomous driving mode, assisted driving mode, partially autonomous driving mode, conditional autonomous driving mode, highly autonomous driving mode, and fully autonomous driving mode. The 2019 Audi A8 has reached the L3 level and is the highest level of autonomous driving in commercial production. The Audi A8 is equipped with the Traffic Jam Pilot system, which allows the car to reach the Partial Autopilot level, i.e., when the speed is less than or equal to 60 km per hour; the user can activate the Autopilot function in congested road conditions. For the Audi A8, Germany has amended its current law to establish 20 ethical guidelines for autonomous and networked vehicles, specifically for cars in autonomous driving mode. China still needs to change the relevant road traffic safety laws, and future changes to road traffic laws are a pertinent issue for self-driving cars.

Another critical issue in the context of autonomous driving is the question of criminal and civil liability after a traffic accident. In terms of criminal law, relevant to this is the'97 Criminal Code regarding the determination of liability for traffic accidents. Two related crimes are the crime of traffic accidents and the crime of dangerous driving. From the current criminal law provisions, dangerous driving refers to driving a motor vehicle in a chase race and the circumstances are wicked or driving a car while intoxicated. The perpetrator must be a natural person who can be criminally responsible, so this crime should not be very relevant to automatic driving. The most pertinent to automated driving is the traffic accident crime, which is mainly manifested in the form of violation of transportation management regulations, thus causing a major accident, resulting in severe injury or death, or causing significant damage to public or private property. From the viewpoint of the current elements of the traffic accident crime, it is impossible to use the self-driving car's manufacturer or software provider as the crime's subject. It is also unreasonable to directly apply the existing criminal law to criminalize the driver in case of a severe accident caused by the car's failure in self-driving mode. There is a big difference. That is to say, when a self-driving vehicle causes casualties, it is difficult to use the existing criminal law to attribute responsibility to the relevant institutions. It is also worth exploring whether they can become the subject of accountability. Another difficulty is that if someone hacks into an autonomous vehicle and causes injury or death to more than one person, who should be held criminally liable?

In civil law, the most relevant aspect of traffic accidents is the law of tort liability and the question of who is liable for compensation. If the problem is due to the quality of the self-driving system itself, then the penalty may be borne more by the car manufacturer or software provider. Still, there will also be problems with the division of the burden of proof, such as reversing the burden of proof, who should bear the responsibility, and so on.

The ethically relevant issues of autonomous driving focus on the tram conundrum, privacy, and unemployment issues.[11] The tram dilemma is a classic proposition previously discussed in philosophy. When a tram is running on a high-speed track and suddenly finds five people in front of it on the trail, there is no way for the tram to stop. The driver has only one choice: a fork in the road next to him, and there is only one person in the division, and he can only choose between the two. So, should the driver continue on the original path or turn onto the next off-ramp? The difficulty here is that the previous philosophical dilemma has yet to have an overwhelmingly accepted answer because it is a choice dilemma. Those who support the sacrifice of one tend to have a utilitarian, holistic view. Those who support traveling on the original path think the one person hit on the off-ramp is a very innocent outsider. One wonders how self-driving systems will make this choice when self-driving cars traveling on the highway may face a similar dilemma to a tram. This requires that the algorithm engineer behind self-driving must understand the ethical implications of both choices, and then the algorithm he coded is likely to be flawed. Giving either of these two outcomes would only be acceptable to some people.

This choice also involves algorithms to make decisions in emergencies. Section 16 of the German road traffic law amendment states that manufacturers and software

[11] Dogan, Costantini and Le Boennec [10], pp. 215–233.

providers cannot require human intervention in emergencies because it is unfair to people. The adjustment of this law reflects the principle of human-centeredness, which then requires the service provider to solve this dilemma in terms of algorithms rather than re-giving the choice to humans in times of emergency. And emergencies require the intervention of human drivers, which may cause much trouble in the overall safety because human beings are in a particular state, their ability to deal with the problem is different. In an emergency, this transfer is a disaster for the elderly or other vulnerable groups who have little capacity to respond to emergencies.

Privacy is a major ethical issue that all smart devices will face in the future because the end of intelligent machines will become a vast platform; they will collect many data from people and exchange them between people, so how to keep this data will be a significant ethical problem. The current emphasis on the concept of v2x is to treat the car as a platform, interacting with many aspects; the car gathers a large amount of data, then how to protect these data is better for the people rather than violate the interests of people will be the most concerned about the issue.

Unemployment is also a significant concern, as the gradual implementation of autonomous driving will inevitably hit people in the traditional driving industry. This number is imposing, for example, the Didi Tax's drop in 2017 of 21 million people driving formal employment. However, in the process of autonomous driving gradually landing, these people may lose their jobs again. And autonomous driving affects small passenger cars, trucks, and other driving jobs. All are issues that we need to consider carefully in the future.

Therefore, the implementation of autonomous driving is a system engineering; we can not only consider the technology but should be considered a variety of aspects, such as legal, ethical, and even people's psychology, to ensure that autonomous driving can improve people's living standards and their travel conditions, rather than creating more trouble for people and causing more and more severe problems.

2.4 The Three Major Components of an Intelligent Society

What is an intelligent society? This concept is closely related to smart cities and smart communities. Cities are composed of systems for organization, government, transportation, communication, water, and energy. These systems are not fragmented but interconnected collaboratively, and the city itself is a macro system made up of these systems. Building a smart city means promoting scientific collaboration and efficient operation of the above methods and the need to achieve sustainable urban development.

An innovative community is a new concept and model of community management. Creative community refers to the integrated application of a new generation of information technology, such as the Internet of Things, cloud computing, and mobile Internet, to provide community residents with a modern and intelligent living environment that is safe, comfortable, and convenient, thus forming a new form of community management based on information and intelligent services. In short, intelligent communities based on the Internet of Things, cloud computing, and other high-tech are a new development stage of community governance and constitute smart cities' grassroots cells.

However, the concepts of a smart city and intelligent community both limit themselves to a specific field, so I propose the idea of an "intelligent society," trying to combine the community, city, and other governmental units and consider the whole society as a governance object in a holistic way. The intelligent society is a new social form based on big data, blockchain, and artificial intelligence. It is the future target society after the agricultural, industrial, and information societies.

The author believes that an intelligent society requires the following three major constituent elements to achieve:

First, intelligent related technologies. These technologies include two main categories, the first of which is intelligent technology. Innovative technology aims to realize the intelligence of things, making many electronic products currently used by people intelligent, friendly, and more convenient. The second category is blockchain technology. Through blockchain technology, people can realize the intelligence of relationships. When, however, this refers to the higher level of blockchain technology. Specifically, blockchain can be divided into three groups: blockchain 1.0 is called digital assets, which is a digital currency with Bitcoin and other cores; blockchain 2.0 is represented by smart contracts, which are currently mainly contract systems implemented based on technologies such as Ether; and the goal of blockchain 3.0 is an intelligent society. In the future, there will be many competent bodies in human society, and the relationship between human and intelligent bodies and intellectual bodies and intelligent bodies will become very complex. The social relationship established by blockchain technology, intelligent contract technology, is a new type of intelligent social relationship. Many researchers divide the two technologies of artificial intelligence and blockchain, while artificial intelligence and blockchain are closely related technologies. The two will form a merging stream in the future and jointly promote the construction of an intelligent society.

Second, thinking intelligently. The technologies that will be needed to build an intelligent society are new to most of the general public. Therefore, we need to integrate these technologies into the consideration of the public so that these technologies are precipitated in the behavior of the public. This requires us to transform this new thinking into people's habits through learning, communication, and social interaction in a flexible way, forming new products and behavior patterns, and finally, forming intelligent thinking. Intelligent thinking contains the following: first, digital thinking. It requires us to understand the existence of society from the perspective of numbers. The digital review also requires us to develop the logic and habits of numbers, using quantitative methods to look at things, analyze problems, and more accurately grasp the laws of problem development; second, intelligent technology thinking. Intelligent thinking requires us to use intelligent thinking to understand things. When a behavior occurs repeatedly and in large numbers, we need to consider whether we can solve such repetitive problems by designing intelligent programs; third, blockchain thinking. Through blockchain technology, fixed social relationships can be presented contractually, and the reproduction of these social relationships can be guaranteed at a meager cost.

Third, the related ethical design. The construction of a new intelligent society requires technical support, and algorithms will become increasingly important. When a large number of decisions need to be made by algorithms, we need to think about the logical structure and ethical implications behind the algorithms. On the one hand, the intelligent society is highly technology-dependent. Still, on the other hand, we need to take a calm look at the social meaning and connotation behind technology. While enjoying the benefits of technological development, we must develop corresponding social tools to manage the social risks and ethical conflicts that technology may give rise to. In this way, while developing technology, we can use it to solve problems instead of adding new social issues. The theme shown in the author's previous book, "Artificial Intelligence: Taming the Seviathan," is the hope that through a series of institutional and ethical designs, we can more fully grasp the new changes brought about by new technologies so that we can better serve social justice.

The intelligent society of the future is highly autonomous. An essential aspect of social governance is strengthening the community's autonomy. Countries that build innovative organizations early on can guide technology development more autonomously through suitable administrative measures and institutional design and encourage people to improve their independence. Take cars as an example; it is a fundamental social norm for cars to give way to pedestrians. This norm can be promoted through camera monitoring and government policy guidance in the early formation. Still, through a long period of advice, people can gradually form the habit of complying with such norms. In building an intelligent society, we need to guarantee the establishment and implementation of standards through state coercive power. Still, in the continuous development of an intelligent society, the mechanism of self-operation of the community will gradually develop, thus forming a mature and stable developed ideal society. In such a social form, every citizen can obey the law, have good moral cultivation, and the community distribution is close to fairness and justice. Every citizen can be expected to enjoy a comfortable and high-quality happy life, which is the ideal state of the future intelligent society (Fig. 2.3).

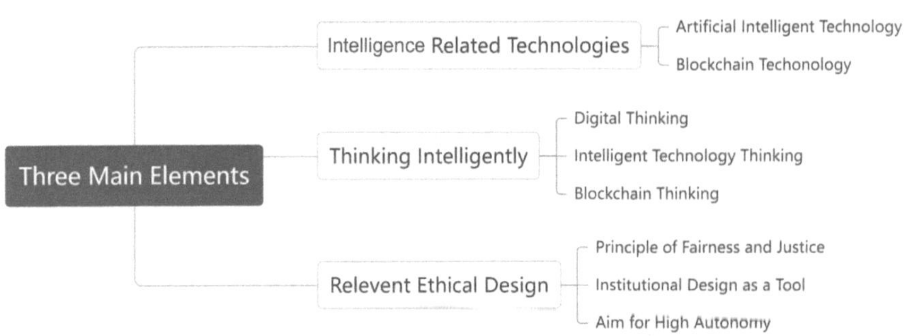

Fig. 2.3 The three main elements of intelligent society

References

1. Ginsberg J, Mohebbi MH et al (2009) Detecting influenza epidemics using search engine query data. Nature 457:1012–1014
2. Bostrom N (2014) Superintelligence: paths, dangers, strategies. Oxford University Press, Oxford, p 52
3. Velarde G (2019) Artificial intelligence and its impact on the fourth industrial revolution: a review. Int J Artif Intell Appl (IJAIA) 10(6):41–48
4. Hosny A, Parmar C, Quackenbush J et al (2018) Artificial intelligence in radiology. Nat Rev Cancer 18:500–510
5. Goksel N, Bozkurt A (2019) Artificial intelligence in education: current insights and future perspectives. In: Sisman-Ugur S, Kurubacak G (eds) Handbook of research on learning in the age of transhumanism. IGI Global, Hershey, pp 224–236
6. Carabantes M (2020) Black-box artificial intelligence: an epistemological and critical analysis. AI Soc 35:309–317
7. Shank DB, Gott A (2020) Exposed by AIs! People personally witness artificial intelligence exposing personal information and exposing people to undesirable content. Int J Hum-Comput Interact 36:1636–1645
8. Caldwell M, Andrews JTA, Tanay T, Griffin LD (2020) AI-enabled future crime. Crime Sci 9(14)
9. Taylor PJ, Dargahi T, Dehghantanha A et al (2020) A systematic literature review of blockchain cyber security. Digit Commun Netw 6(2):147–156
10. Dogan E, Costantini F, Le Boennec R (2020) Chapter Nine—Ethical issues concerning automated vehicles and their implications for transport. Adv Transp Policy Plan 5:215–233

Chapter 3
The Future of Global Governance of Artificial Intelligence

The current global governance mechanism, which is still dominated by sovereign states, has two core features: hegemonic logic and conflict logic. The application of artificial intelligence in global governance will not naturally lead to the dissolution of the hegemonic logic. The development of artificial intelligence does not necessarily lead to a decline in the dominant power of Western countries such as the United States, while developing countries face both opportunities and challenges. Therefore, global good intelligence means that major countries should form a global consensus on the development of strong AI and super AI, adopt a global consultation mechanism to regulate the rhythm of the development of intelligence, take explainable and safe AI as the future development direction, promote the solution of historical problems in developing countries through intelligence, and strike a balance between global governance and national governance. The concept of global good intelligence is based on the "three new principles of robotics", which include: first, AI is always an aid; second, the proportion of human decision-making should not be lower than the golden ratio; third, humans should always grasp the rhythm of AI development, and be ready to pause or slow down. At the same time, the application of AI in global governance can hardly lead directly to the decline of conflict logic. In the long run, the development of AI will help people to improve understanding, but in the short term direct and frequent contact may increase new conflicts. At the same time, AI technologies can help people on the move to understand other cultures, but the anti-immigrant wave exacerbated by global unemployment is not conducive to the free global movement of people around the world. Based on this, global collective intelligence requires the global society to form synergy in man–machine intelligence, multinational intelligence and multi-actor intelligence. Global good intelligence and global collective intelligence can become the target value and process value of artificial intelligence application in global governance in the future.

Q. Gao and J. Zhang, *Artificial Intelligence Governance and the Blockchain Revolution*, Artificial Intelligence and the Rule of Law,
https://doi.org/10.1007/978-981-99-9211-9_3

3.1 The Current Structural Dilemma of Global Governance

Along with the increasing globalization process, global governance has gradually become the consensus of most countries in the international community; however, global governance is now facing some structural dilemmas. Meanwhile, as a leading technology in the Fourth Industrial Revolution, artificial intelligence is developing rapidly and profoundly changing human life and world politics. The central question discussed in this section is whether the development of AI will contribute to the solution of structural problems of global governance. The chapter first summarizes the two major sticking points of current global governance: the logic of hegemony and the logic of conflict. Then two analytical hypotheses are proposed: first, the development of AI helps to reduce the hegemonic logic, and second, the development of AI helps to reduce the conflict logic. For the depth of the discussion, we will decompose these two major hypotheses into several sub-hypotheses and examines the impact of AI development on each of these factors. After discussing these issues empirically, we will lead the study back to theoretical research and proposes two ideas—global good intelligence and global collective intelligence—in response to the empirical analysis, and discusses the possibility of these two theories for eliminating the logic of hegemony and the logic of conflict.

Despite the strong emphasis on polycentric as well as global mechanisms at the beginning of the concept of global governance, in practice, global governance is still concentrated on state-driven global governance. On the whole, non-state mechanisms have not yet played an adequate and effective role. These non-state mechanisms include global mechanisms, regional mechanisms, and global NGOs.

First, the role played by the United Nations and other international organizations still faces many limitations. Although international organizations such as the United Nations, the World Bank, International Monetary Fund, and World Trade Organization have played a role in global governance, their advancement in various fields still faces many constraints. For example, a study by Tetsuro Iji of Toyo University argues that the UN's function as a mediator in international conflicts only has some room for those conflicts in which the interests of major powers are not involved.[1] Again, the World Bank seeks to promote the economic development of developing countries through the establishment and guidance of a series of development norms. To avoid opposition from developing countries to the World Bank's development model output, the Economic Development Institute and the World Bank Institute, two of the World Bank's affiliates, persuade member countries to accept these development norms through what they define as best practices and case studies.[2] However, the translation of these development norms into the economic development practices of developing countries is another matter. For example, a study by Richard Mansbach, an American political scientist and Ellen Pirro, a professor of political science at Iowa State University, examined the impact of institutions such as the World Bank and the International Monetary Fund on the European recovery after the 2008 financial

[1] Iji [1].

[2] Bazbauers [2].

crisis and concluded that these institutions had little impact on the recovery itself.[3] In addition, the World Trade Organization is under pressure to reconfigure, and global climate governance negotiations are at an impasse.

Second, regional governance is now also facing new and notable difficulties. Arie Kacowicz, a professor of international relations at Georgetown University, summarizes the relationship between regional and global governance as irrelevance, conflict, cooperation, and harmony, and argues that the EU, as a pioneer in regional governance practices, has a model effect on global governance.[4] Although the EU has made great achievements, it also faces many challenges to its development in the context of the Brexit event and the European economic downturn, which makes regional governance as an auxiliary form of global governance also come under some questions.[5] Other regions such as North America, East Asia, Latin America, Africa and other regional mechanisms have also not seen greater development.

Once again, global NGOs, although very active, do not have the capacity to drive substantive change in global governance. In the 1990s, when the theory of global governance was first proposed, Western academics identified civil society organizations as a key driving force for global governance reform, but this is now seen as highly unrealistic. CSOs are more of a fragmented force that has a direct role in promoting relatively micro and focused issues (typically in areas such as global health governance), but the solution to these micro issues hardly touches the essence of global governance. As a result, CSOs are unable to promote the reform of the entire global governance system, nor can they propose comprehensive solutions to many structural and fundamental problems.

Overall, the biggest problem of global governance at present is that the global community lacks mechanism, a systematic framework and a roadmap for the advancement of global issues. Global issues are problems faced by the international community that require global thinking, including specifically regional conflicts, North–South relations, ecological issues, environmental pollution, terrorism, transnational crime, population explosion, and resource shortages. Although these issues are relatively diverse, after specific analysis, they can be grouped more broadly into the following categories:

The first category is developmental issues, i.e., problems resulting from the underdevelopment of developing countries. Some of the major global problems of today, such as population explosion and epidemics of infectious diseases, fall into this category. The root causes of global problems such as transnational crime and international terrorism are also closely related to these problems. If underdeveloped countries do not make significant advances, developmental issues can be considered as one of the important root causes of global problems.

The second category is the hegemonic problem, that is, the developed countries abuse their power in the global order and cause the development problems of the underdeveloped countries to worsen, or their unwillingness to do some self-restraint

---

[3] Mansbach and Pirro [3].

[4] Kacowicz [4].

[5] Hobolt [5].

in the international mechanism to promote the development of the global mechanism. The formation of a global mechanism itself requires some concessions from all countries concerned. However, an important trend is that developed countries are still reluctant to make concessions on issues such as climate, trade, and North–South issues, despite the fact that most of the conditions are significantly better than those of developing countries.

The third category is comprehensive problems, i.e. complex problems caused by a combination of both the underdevelopment of developing countries themselves and hegemonic acts such as the abuse of force by some developed countries. Regional conflicts fall mainly into this category.

If these problems remain unresolved or accumulated for a long time, they will lead to a global governance crisis.[6] Further, there are two important features of the current global governance mechanism, which is still dominated by sovereign states, as follows:

First, the logic of hegemonic. Global governance is still dominated by the U.S.-centered Western hegemony. The goal of developed Western countries is often to maintain Western dominance in the world at large (especially in areas such as finance and technology). As a result, Western countries will use many previously developed mechanisms of dominance to maintain this competitive advantage.[7] This is one of the major causes of chronic poverty in the non-Western world. In recent years, while some emerging developing countries have achieved greater development, they also have the will to assume a more important role in the international community and push for reform of the existing global governance system, but these positive actions have been ignored or even resisted by the developed Western countries. Some anti-globalization phenomena that have emerged in the West in recent years are the manifestations of this hegemonic logic of governance. When Western countries gain huge benefits from the globalization process, they will greatly promote this process. However, when more and more developing countries join this process and slowly enter some central areas, some developed countries may take measures to oppose or prevent the further development of globalization.

Second, the logic of conflict. The persistence of the current structural problems in the field of global governance is closely related to the conflict-based thinking in the mainstream Western culture. This conflict-based thinking originates from the assumption of human evil in Christian culture, which sees relations between actors as often conflicting.[8] For example, realism is the most important theoretical school in Western international relations theory, and at the same time realism is a typical representative of conflicting values.[9] Although Western society has shown a certain secularization in many aspects, Christian culture remains the most important ideological

[6] Honghua [6].

[7] Van der Pijl [7].

[8] Qiqi [8].

[9] Hans J. Morgentha, a representative of traditional realism, has made the struggle for power the main content of international politics, and Morgenthau's theory derives from Schmitt's philosophy of conflictiveness. See Pichler [9].

core within the developed Western world. As a monotheistic religion, the underlying implication of Christian doctrine is that only pagans who convert to Christianity can truly gain their recognition. As a result, other civilizations that maintain their distinctiveness are often considered heretical or alternative by Christian civilization. For example, the European left-wing thinker Slavoj Žižek has a wonderful comment on the exclusivist character of Christianity: "In other 'particularistic' religions (and even in Islam, in spite of its global expansionism), there is a place for others, they are tolerated, even if they are condescendingly looked upon. The Christian motto 'All men are brothers', however, means ALSO that 'those who are not my brothers ARE NOT MAN.' ...... while Christian universalism tendentially excludes non-believers from the very universality of humankind."[10] The American historian Arnold Toynbee has also commented on this exclusionary nature of Western culture and Christianity: "In the spiritual realm, Western rationalists and Western Christians share a condescending attitude" that "scorns all other religions".[11] This tradition makes it difficult for Christian civilization to accept and identify psychologically with other civilizations. This is the cultural root of the difficulty for developed Western countries to accept developing countries. Therefore, in a sense, the Christian civilization represented by the developed Western countries cannot accept and tolerate the progress and development of other civilizations from their ideological roots, which is an important ideological and cultural background for the persistence of poverty problems and regional conflicts. Moreover, in the context of global mobility, the influx of foreign populations to the Western world has, to some extent, intensified the conflict between different cultures. The inability of non-Western immigrants to truly integrate into Western society can also be traced back to cultural differences. In short, this logic of conflict is essentially a logic of exclusion.

This logic of conflict is concentrated in the discussion of Western scholars on global governance. For example, Matthew Stephen, a researcher at the Center for Social Science Research Center Berlin, argues that the impact of the rise of emerging countries on global governance can be summarized in the following six points: (1) global governance is here to stay, but is increasingly contested; (2) a struggle is ensuing for leadership and privilege within global governance; (3) the liberal social purpose of global governance is taking a back seat; (4) existing multilateral institutions are facing increased deadlock; (5) informalization is likely to increase; and (6) global governance is becoming more fragmented.[12] Stephen's last three judgments are largely correct in seeing the problems currently facing global governance. However, these three problems are not caused by the rise of emerging countries, but rather by the difficulties of the traditional global governance system itself. The first three judgments, on the other hand, are strongly characterized by hegemonic and conflict logic. The first and the second judgments both view the relationship between emerging and established countries in global governance as a zero-sum game. The third judgment implicitly laments the ideological decline of Western

[10] Zizek [10].

[11] For more details, see Toynbee [11].

[12] . Stephen [12].

countries. Stephen's conclusion is similarly distinctly conflictual: "In this respect, the new global governance is likely to be more institutionally diverse and polycentric, with more signs of overt conflict compared to the recent past as emerging powers are able to afford to openly disagree with established powers. In any case, 'problem solving' and 'cooperation' cannot adequately describe the central dynamics of the new global governance—politics is back."[13]

Again, Craig Lerner, a professor of law at George Mason University, argues that the idea of global governance is a modern metaphor for the Tower of Babel. People emphasize the significance and necessity of global governance by describing future risks such as climate change, nuclear war, artificial intelligence, nanotechnology, etc. Lerner argues that the idea of global governance is wrong in light of the Bible's description of the Tower of Babel. Implicit in the Tower of Babel metaphor itself is the notion that the division of humanity into separate spheres will have beneficial consequences. In Lerner's view, in addition to the feasibility issues, global governance solutions exaggerate the benefits and underestimate the costs of cooperation, while maintaining separate, even competing, nation-state approaches often has substantial advantages.[14] Explaining world politics from the perspective of Christian culture, Lerner's viewpoint is in a logical contradiction. Understood in a more abstract sense, the essence of this logic of conflict is the alienation of man. According to Karl Marx, under the influence of the capitalist system, the subject-object relationship is alienated. The result is that the relationship between man and nature is not unified, but rather antagonistic. And, this same antagonism that is found in Western ideology spreads within human society. Marx points out that "Whenever real, corporeal man, man with his feet firmly on the solid ground, man exhaling and inhaling all the forces of nature, posits his real, objective essential powers as alien objects by his externalisation, it is not the act of positing which is the subject in this process."[15] In other words, when the relationship between man and nature, between man and man, is alienated, it is not only the object but also the subject that is threatened, and it is not only capitalist society itself that is threatened, but the whole human society and the global governance system (see Fig. 3.1).

3.2 Analyzing the Hypothesis: Can AI Change the Two Crux Logics of Global Governance

Human society is experiencing the Fourth Industrial Revolution. The First Industrial Revolution began in the 1860s, centered on the steam engine revolution, and was mainly defined by Britain. The Second Industrial Revolution began in the 1860s, centered on the electrical and internal combustion engine revolutions, and

[13] Stephen [12, p. 498].

[14] Lerner [13].

[15] *Karl Marx, Frederick Engels: Complete Works*, Vol. 3, Beijing: People's Publishing House, 2002, p. 324. (in Chinese).

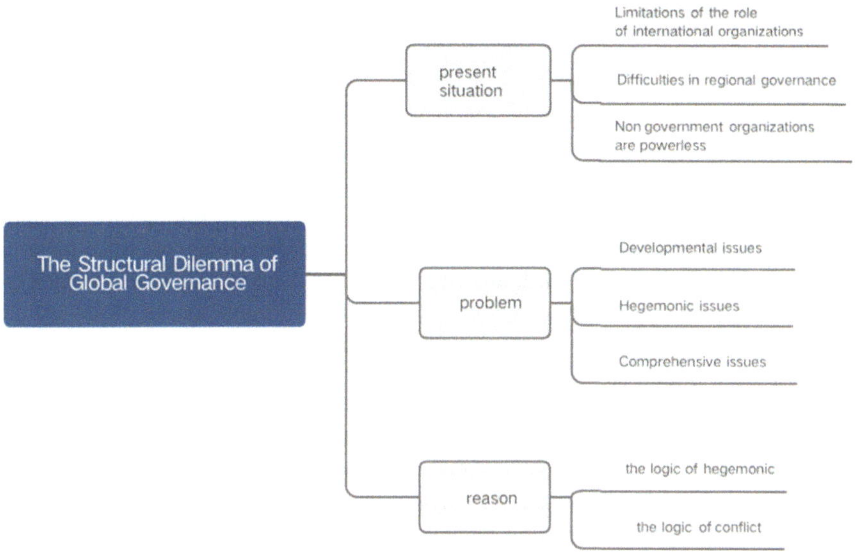

Fig. 3.1 The dilemma and root causes of global governance

was mainly defined by Germany and the United States. The Third Industrial Revolution, which began in the 1940s, centered on the computer and information revolutions and was mainly defined by the United States. The Fourth Industrial Revolution is currently taking place, and most studies consider its leading technology to be artificial intelligence.

The definition of artificial intelligence has been a theoretical challenge. Stuart Russell, professor of artificial intelligence at the University of California, Berkeley, and Peter Norvig, AI scientist at Google, have summarized the current definition of AI into four categories: first, acting like a human, i.e., the Turing Test pathway; second, thinking like a human, i.e., the cognitive modeling pathway; third, thinking rationally, i.e., The way of thinking rules; fourth, act rationally, i.e., the way to act rationally.[16] In short, AI is the simulation of human intelligence, i.e., the use of machines to simulate human intelligence and to assist people in accomplishing some tasks or decisions. The landmark event in the origin of artificial intelligence was the Dartmouth Conference in 1956. Ten prominent information scientists such as John Mckarthy, Marvin Minsky, Claunde Shannon, and Herbert Simon attended this conference and laid down some basic frameworks for the subsequent development of artificial intelligence. Since then, AI has gone through three waves. The first wave, which ran from 1956 to the late 1960s, centered on the use of reasoning by computers to solve specific problems, such as using machines to prove rules. Despite the great achievements of this period, this reasoning approach was helpless for complex real-world problems. When this weakness became increasingly apparent, the first wave of AI boom declined. Thus, in the 1970s, AI research entered its first severe winter. The

[16] Russell and Norvig [14].

second wave emerged in the 1980s and centered on knowledge engineering, which is the use of expert systems to solve real-world problems. However, the knowledge engineering approach has its limitations. After left in 1995, AI entered the winter again. The third wave emerged after 2000, especially after 2006, when AI approaches based on deep learning became prevalent along with the Internet boom and the emergence of massive amounts of data.[17] There are three main reasons for this rise: first, the amount of data has increased dramatically due to the proliferation of sensors; second, the computing power of computers has also increased significantly compared to what it was before; and third, algorithms such as deep learning, represented by deep neural networks, have shown excellent results in applications. What we are currently experiencing is the rising period of the third wave of artificial intelligence.

The Fourth Industrial Revolution was created on the basis of the Third Industrial Revolution. The development of artificial intelligence relies on important foundations such as computility, algorithms and data, which are built on the basis of information technology. In a sense, AI is also seen as a new development or new phase of information technology. Traditionally, AI also emerged initially as a sub-discipline under the broad category of information science such as computer science or automation. The Fourth Industrial Revolution is also associated with a number of other technologies, such as the Internet of Things, virtual reality, blockchain, etc. However, all of these technologies are closely linked to artificial intelligence in some way. It is called the intelligent revolution because the impact of AI on various fields is huge. For example, autonomous driving, aided by AI technology, could have a profound impact on traditional transportation. Likewise, AI has profound impacts on various fields such as healthcare, education, law, finance, and media.

Because AI has been developed in the United States for more than six decades, scientific achievements related to AI and political science and public administration in the United States predate the third wave. For example, Richard Kimber, from the United Kingdom, used the ID3 algorithm in artificial intelligence to build neural networks and predict the degree of democracy that transitional countries might achieve in the future based on socioeconomic data from 147 countries.[18] Thomas Barth, a professor at the University of North Carolina, and Eddy Arnold of the University of Memphis discuss some of the dilemmas related to administrative discretion that may arise from technological developments in artificial intelligence and summarize the benefits and dangers of artificial intelligence for the field of public administration.[19] Gavan Duffy, associate professor of political science at Syracuse University, and Seth Tucker, professor of global security at Syracuse University, also discuss the promise of AI applications in political science and the issues facing the future.[20] However, direct references on the impact of AI on global governance are scarce. Results on the impact of the third wave of AI on global governance and international relations are also very limited in the Western literature and the most relevant is the

[17] Matsuo [15].

[18] Kimber [16].

[19] Barth and Arnold [17].

[20] Duffy and Tucker [18].

study of the impact of big data on global governance. Chinese scholars have noticed the impact of big data on the study of international relations.[21] At the same time, relevant studies by Chinese scholars on AI and international relations have begun to appear sporadically.[22]

Two limiting features of current global governance mechanisms were discussed above: first, the logic of hegemonic, and second, the logic of conflict. Therefore, the following is organized around the following two hypotheses: first, the development of AI helps to reduce the hegemonic logic (hypothesis 1). Second, the development of AI facilitates the reduction of conflict logic (hypothesis 2).

The first hypothesis of this chapter is that the development of AI helps to reduce the logic of hegemonic. This hypothesis can be divided into three sub-hypotheses: 1. The development of AI makes the dominant power of western countries such as the United States decrease (hypothesis 1-a). 2. Developing countries have better development opportunities in the intelligent revolution (hypothesis 1-b). 3. International rules on AI develop in the direction of favoring developing countries (hypothesis 1-c).

The second hypothesis of this paper is that the development of AI facilitates the reduction of the logic of conflict. This hypothesis can be divided into two sub-hypotheses: 1.The development of AI helps to increase the understanding between different groups (hypothesis 2-a). 2. The development of AI contributes to the global free movement of populations (hypothesis 2-b).

3.3 Reinvention or Dissolution? The Role of Artificial Intelligence on the Logic of Hegemonic

With regard to hypothesis 1-a, "The development of AI makes the dominant power of western countries such as the United States decrease", the following facts can be observed to be relevant:

First, the U.S. superpower in artificial intelligence is very obvious, while the gap between other developed Western countries and the U.S. is widening. Since the intelligent revolution is an extension of the information technology revolution, the United States has super power in the competition. This super power is mainly manifested in the following points: First, the leading companies of intelligent computing hardware are almost all American companies. For example, Nvidia has increased influence because its GPU has advantages in AI computing, Intel is also fully laying out AI chips, Qualcomm has advantages in smartphone chips, and Google is developing smart dedicated chips such as TPU. Second, U.S. companies and universities are very strong in AI algorithmic frameworks and their ecology. The most popular algorithmic frameworks for artificial intelligence such as Google's TensorFlow, IBM's

[21] See Jian [19]; Haixia [20]; Qingling [21].

[22] The author has organized a set of articles on the topic of AI in the journal International Insight. See Qiqi [22]; Weiguang and Jing [23]; Qingling [24]; Chuanying and Mallery [25]. Other related articles include Shuai [26].

SystemML, UC Berkeley's Caffe, and Facebook's Torch are almost all led by U.S. companies or research institutions. Third, in specific application scenarios, U.S. companies have a clear advantage. For example, in the field of autonomous driving, Google's Waymo and Tesla, among others, are the most leading companies. Then again, in the field of special robots, Boston Dynamics' technology is also in the leading position. Fourth, the United States also has an advantage in certain original innovations. For example, OpenAI, backed by Silicon Valley giants such as Tesla CEO Elon Musk and PayPal founder Peter Thiel, is doing some disruptive research, such as building "universal" robots and chatbots that use natural language. The U.S. superpower is in contrast to other developed Western nations. The U.S. superpower has also opened up a gap with other developed Western nations. Although Germany and Japan are also traditional industrial powers, but the gap between the two countries and the United States in artificial intelligence is still significant. In addition, among the traditional developed countries, the United Kingdom, Canada, Israel, etc. also have a good foundation in the field of artificial intelligence, but all have a large gap with the United States.[23]

Second, despite China's rapid growth in AI in recent years, uncertainty about the future is still evident. China is a new entrant in the field of artificial intelligence. There are several reasons why China has developed rapidly in recent years. First, China has a solid foundation in the information technology sector. China has caught the wave of mobile Internet development in the last two decades, and the gap with developed countries has been narrowed greatly. In the development process from 2 to 3G, Qualcomm was the absolute dominant player. In the process of 3G to 4G, China then became one of the game changers. The TD-LTE standard led by China occupies an important position in 4G. At the same time, China's cell phone brands such as Huawei, OPPO, Vivo and Xiaomi developed rapidly and gradually competing with giants such as Apple and Samsung. By the time 5G landed, China had basically kept pace with developed countries, and was even ahead in some technologies. Second, the active guidance of the national strategy. The State Council of the People People's Republic of China issued the "New Generation of Artificial Intelligence Development Plan" in July 2017, followed by the Ministry of Industry and Information Technology of the People's Republic of China, who issued "Three-Year Action Plan for Promoting the Development of a New Generation of Artificial Intelligence Industry (2018–2020)" in December 2017, while the Ministry of Education of the People's Republic of China issued the "Action Plan for Artificial Intelligence Innovation in Higher Education" in April 2018. To ensure the landing and implementation of this national strategy, various provinces and cities in China have also formulated relevant promotion plans. Moreover, Political Bureau of the CPC Central Committee conducted a collective study on the current status and trends of AI development in October 2018, which will also create an important impetus for the next development of AI. Third, China's large market and neat consumer groups lay the foundation for AI to land. China's huge amount of data come from its large population and high level of informationization, which are the basis for the rapid implementation of AI

[23] Qiqi [27].

in various scenarios. Fourth, China's economic take-off in recent years has led to active private capital and therefore a rapid increase in investment in AI.

China's achievements in the field of AI have grown rapidly in recent years. For example, "the China AI Development Report 2018" released by Tsinghua University states that China ranks first in the world in terms of the total number of papers and the number of highly cited papers. China is also slightly ahead of the United States and Japan in terms of the number of patents in AI.[24] In the field of AI, China's biggest feature by far is the ability to turn cutting-edge applications into products relatively quickly. In recent years, China's rapid development has been concentrated in the application areas of visual recognition and speech recognition. For example, on the basis of visual recognition technology, Hikvision, Dahua and other companies have grown rapidly into the world's top-ranked intelligent security companies in terms of scale and technology. In the field of chips, some application-based chips have made great progress, such as HUAWEI Kirin, Cambrian's NPU, Horizon's BPU, Alibaba DAMO Academy's Ali-NPU under research, and so on. Of course, China still has obvious shortcomings. For example, there is still a great lack of intelligent basic chips, algorithm frameworks and ecology, while some basic theories and native innovations are also relatively insufficient.

Third, developed countries, especially the United States, wish to lock in their technological advantage over developing countries in the field of artificial intelligence. On the issue of technological progress, developed countries seek to use various mechanisms to freeze technological advantage, which has largely led to many developing countries being confined to low-value output jobs in relatively low-end industries. Since the value added in these low-end industries is very low, workers in developing countries remain poor despite their relatively high labor effort. In the long run, developing countries are trapped in a vicious circle: because of their low economic level, they are unable to improve the education level of their citizens, which in turn prevents them from improving their R&D capabilities in cutting-edge science and technology, and makes it difficult for influential industries and companies to emerge to support the whole country. In this way, developing countries are trapped in a strange circle and cycle of non-development. To solve the most fundamental development problem in global governance, the most important measure should be to improve the technological level of these countries.

However, the very critical issue at present is that the developed Western countries are not willing to transfer frontier science and technology knowledge such as artificial intelligence to developing countries. At the same time, when some emerging countries in developing countries make breakthroughs in certain frontier areas, Western developed countries will also use foreign investment reviews, export controls, and restrictions on scientific and technological exchanges and personnel exchanges to prevent the scientific and technological progress of emerging countries. The developed countries hope to keep advanced technologies in their hands for a long time, so as to achieve the purpose of dominating the new scientific and technological revolution in the long term by hindering the diffusion of technologies. Fundamentally, the

---

[24] Hao [28].

developed Western countries still have not got rid of the traditional imperialist and colonialist thinking. The traditional colonization method of the developed Western countries is geographical colonization. However, along with the development of national liberation movements, especially after World War II, geographic colonization became less and less possible, and technological colonization became its main alternative form. Western countries prevented developing countries from entering many technological fields through the control of patented technologies, thus freezing the gap between developed and developing countries. Considering from this perspective, patented technologies are largely in the service of technological hegemony. Those super-major companies keep acquiring patents through mergers and defeating their rivals through litigation, so intellectual property serves the hegemonic strategies of Western countries to a certain extent. In the era of artificial intelligence, the monopoly of supermajor companies on core technologies will become more prominent. Matsuo Yutaka, an associate professor of artificial intelligence at the University of Tokyo in Japan, has reminded us to be wary of technology monopoly in the AI era. If certain algorithms in AI are controlled by specific companies and operated "under the table," it is unfair to other companies and the public. In addition, companies that have a common operating system will have a competitive advantage that will eventually lead to a monopoly in the market.[25] Ariel Ezrachi, a professor of law at Oxford University, and Maurice Stucke, a professor of law at the University of Tennessee, also point out that when some super platform companies extend their tentacles to virtual assistants, the Internet of Things, and smart devices, their data advantage will evolve into a competitive advantage and market power. Thus, the worst-case scenario for the future is that money will be concentrated in a few superplatforms as access to wealth grows.[26]

With respect to hypothesis 1-a, the preliminary conclusion at this point is that the U.S. advantage remains clear and that, despite China's rapid progress, the U.S. is likely to place limits on China's development, which could lead to increased uncertainty about China's development.

With respect to hypothesis 1-b, "Developing countries have better development opportunities in the intelligent revolution", the relevant facts that can be observed include:

First, open-source software and new learning approaches can help develop talent in developing countries. At this stage, many industrial fields related to artificial intelligence have technologies that are open source. Open-source technology means that latecomers can more easily use the research results of their predecessors, thus greatly reducing the time to catch up. In the era of AI, the scarcity of high-level educational resources will continue to decrease, and people will have easier access to knowledge. For example, using the current live streaming technology and short video technology for knowledge learning will produce very good results. There are already some educational products in use, such as the domestic Xueersi Online School. If people in developing countries can give full play to the technological potential of AI,

[25] Matsuo [29].

[26] Ezrachi and Stucke [30].

empower learning through technological ways, and improve the overall education level and national literacy of the country, they can lay a solid foundation of talent for economic development. In essence, developing countries can be empowered by AI technology to gradually improve their internal governance by raising the education level of their nationals, and eventually achieve a state of freedom from poverty.

Second, institutionalization driven by the intelligent revolution helps developing countries overcome the limitations of traditional cultures. Artificial intelligence-related technologies can help solve some traditional problems in the first place. Research by Hans Hansen, management at Copenhagen Business School, and Tony Porter at McMaster University points to cell phones and big data technologies as key in the fight against diseases such as Ebola. Cell phones help people send information about coordination-related efforts, while analysis of call data records helps epidemiologists track the spread of disease.[27] Introducing AI solutions to problem governance in developing countries can go a long way toward reducing the traditional human element of interference. Artificial intelligence can overcome the over-reliance of human intelligence on emotional factors and can help traditional societies transition to a jurisprudence-based society. The backwardness of developing countries is largely cultural, and the machine rationality of AI can help hedge against the influence of traditional culture. For example, many developing countries are plagued by corruption; however, with all transactions recorded through information systems, electronically enabled payments or mobile payments, etc., the traditional opportunities for corruption are reduced. Of course, in this process of institutionalization, developing countries need to find a balance between foreign experience and their own culture. Many of the current international indicator systems are based on the experience of developed Western countries, and once these measurement tools are simply applied to the practices of developing countries, many problems of applicability arise.[28] Therefore, such data- and measurement-based global governance needs to be centered on the practices of developing countries. Research by Jacqueline Best, a professor of political science at the University of Ottawa, has also found that current global governance is increasingly focused on measurement, ranking, and scoring. The heavy use of quantitative methods and assessment techniques in matters of international development has instead become a source of power exerted by certain states. In many cases, these powers reinforce existing asymmetries and diminish the actual potential of these governance practices.[29] In light of this, global governance practices on top of AI technologies likewise have to return to the realities of developing countries.

Third, the advent of the intelligent era may lead to the end of the international division of labor, thus making the gap between developed and developing countries further widen.[30] First, the foundation of world system is the division of labor. According to the "mainstream dependency theory" proposed by Brazilian economist

[27] Hansen and Porter [31].

[28] De Siqueira [32].

[29] Best [33].

[30] Qiqi [34].

Dos Santos, in the 1950s and 1960s, an indivisible world system was formed between Western capitalist countries and developing countries due to the development of monopoly capitalism. Different countries have different resource. Developed countries have advanced technology but lack labor, while developing countries can provide abundant cheap labor. Therefore, developing countries have the opportunity to fully participate in the international division of labor and obtain the basic resources for labor quality improvement, and then further improve their national strength and labor quality as a whole through skills spillover and technological upgrading.[31] For example, the American economist Alice Amsden pointed out that in the 1920s Japan, for example, was able to penetrate the British Lancashire market with its textiles, relying not only on trading companies that could provide raw cotton, modern equipment and integrated production lines, but also on cheap labor. By the 1950s and 1960s, other late-industrializing countries, led by South Korea and China Taiwan, were trying to break into international export markets with the comparative advantage of low wages, and these late-industrializing countries and regions represented a new international economic order.[32]

However, the development of artificial intelligence is likely to destroy this foundation as developed countries gradually realize that the cost of robots is largely lower than that of maintaining human labor. In some developing countries, the cost of labor use is initially low, but in the process of globalization, as labor standards and benefit levels rise in these developing countries, the cost of labor rises. In contrast, the cost of artificial intelligence is more manageable. This reliance on machines to some extent precludes developing countries from participating in a worldwide division of labor. Obviously, if AI technology develops to a higher level in the future, the labor-intensive division of labor undertaken by developing countries in the international division of labor will be gradually replaced, and then it will be difficult for developing countries to participate in the previous world division of labor. As MIT management professors Erik Brynjolfsson and Andrew McAfee outline, "In the long run, automation will probably affect most not workers in the U.S. and other developed countries, but rather low-cost labor as a competitive advantage for developing countries."[33] At the same time, the end of developing country participation could further widen the technological divide between North and South. Barriers constructed by developed countries in new AI technologies will grow higher, and underdeveloped countries will become less likely to enter the heartland through participatory learning. Danah Boyd and Kate Crawford argue that limited access to big data has created a new data divide. Large data companies have absolute monopoly power over data.[34] The access divide is therefore the basic form of the digital divide.[35] Mark Anderjevic, professor of journalism and communication at Monash University, uses the big data divide to describe the asymmetric relationship between those who collect, store and

[31] Santos [35].

[32] Amsden [36].

[33] Brynjolfsson and McAfee [37].

[34] Boyd and Crawford [38].

[35] Zechi et al. [39].

mine large amounts of data and the targets of that data collection. Anderjevic argues that the growth of big data may exacerbate the power imbalance in the digital age.[36] Hansen et al. argue that big data relies on technology to operate globally, obscuring and obfuscating power in new ways. As the actual impact of big data in international affairs continues to expand, effective ways to adjust to these changes in power relations are necessary.[37] Indeed, the new types of inequality resulting from the development of artificial intelligence can be understood as the alienation of artificial intelligence. According to Marx, there is an alienation of science and technology, which is rooted in the "capitalist application of the machine". Marx emphasized that "The contradictions and antagonisms inseparable from the capitalist application of machinery do not exist, they say, because they do not arise out of machinery as such, but out of its capitalist application!"[38] The inequalities caused by the alienation of artificial intelligence will not only appear within capitalist society, but will also spill over into the international system of division of labor.

Therefore, regarding hypothesis 1-b, the current preliminary conclusion is that although developing countries can get some technological dividends and latecomer advantages of AI, since AI itself is a high-end frontier technology, and developing countries lack sufficient talent pool, then the gap between developing and developed countries may become increasingly wide.

With respect to hypothesis 1-c, "international rules on AI develop in the direction of favoring developing countries," the relevant facts that can be observed are:

First, the current global AI-related laws and regulations, policies, and principles are mainly defined by developed Western countries. Developing countries have a very weak voice in them and cannot even participate in the relevant discussions. For example, some large companies in Western developed countries have the dominant power in the definition of these rules. "The Asilomar AI 23 Principles (hereafter referred to as the "The Asilomar AI Principles") are the most influential AI principles in recent years.[39] Although the Asilomar AI Principles claim to be signed by nearly 1,000 experts in the field of AI and robotics, a closer look reveals that the main promoters of the principles are still Western entrepreneurs like Musk. Some of the macro-level elements of the Asilomar AI Principles are fundamentally correct. For example, the emphasis on "beneficial intelligence" in the "research objectives," the emphasis on "Safety of AI systems" in Article 6, and the emphasis on "Failure Transparency" in Article 7, "Judicial Transparency" in Article 8, "Human Values"

---

[36] Anderjevic [40].

[37] Hansen and Porter [31, pp. 31–32].

[38] *Karl Marx, Frederick Engels: Complete Works*, Vol. 44, Beijing: People's Publishing House, 2001, p. 508. (in Chinese).

[39] The Asilomar AI 23 Principles are divided into three categories, with a total of twenty-three articles. The first category is for research issues (with five articles), including the purpose of research, funding policies, culture and competition, and other elements; the second category is for ethical value issues (with thirteen articles), including safety, responsibility, and values in AI development; and the third category is for long-term issues (with five articles), aiming to address the potentially catastrophic risks caused by AI development. Future of Life Institute, *Asilomar AI Principles*, January 3, 2017. https://futureoflife.org/ai-principles/?cn-reloaded=1.

in Article 11, "Shared Benefit" in Article 14, and "Shared Prosperity" in Article 15 are all relatively easy to reach a consensus with the public.

However, there are still problems with the Asilomar AI Principles. First, these principles imply the goal of developing general AI or even super AI. Both Articles 9 and 10 of the Principles emphasize that "Advanced AI systems and Highly autonomous AI systems should be designed", while Article 16 emphasizes that "Humans should choose how and whether to delegate decisions to AI systems, to accomplish human-chosen objectives". Article 17 emphasizes that "The power conferred by control of highly advanced AI systems should respect and improve, rather than subvert, the social and civic processes on which the health of society depends". Article 19 emphasizes that "we should avoid strong assumptions regarding upper limits on future AI capabilities". Article 20 emphasizes that "Advanced AI could represent a profound change in the history of life on Earth, and should be planned for and managed with commensurate care and resources". General AI is the term used in contrast to specialized AI. Specialized AI, also known as Modular AI, has narrow expertise in a specific domain and can learn by doing to improve its performance. In contrast, general AI can apply its knowledge more flexibly to a more abstract and infinite set of problems, including those that require an understanding of meaning and value.[40]

In short, specialized AI is intelligence that is limited to a particular domain, while general AI is intelligence that seeks to achieve comprehensive human-like problem-solving capabilities as a whole. General AI is also often referred to as strong AI, and the next stage in the development of general AI is super AI. If the goal of general AI is to be equivalent to humans, the goal of super AI is to develop intelligence that exceeds human intelligence. Nick Bostrom defines superintelligence as "any intellect that greatly exceeds the cognitive performance of humans in virtually all domains of interest". Bostrom also classifies superintelligence into the following categories: A speed superintelligence, where the system could do what a human does, but faster; A collective superintelligence, where the system is composed of smaller intellects. It is especially good at tasks that can be broken into parts and completed in parallel; and a quality superintelligence can carry out intellectual tasks that humans just can't in practice.[41] Many experts have shown concern about the development of super-artificial intelligence. Russell and Norvig, for example, argue that the success of AI could mean the end of the human race. In the wrong hands, almost any technology has the potential to cause harm. For AI and robotics, our new problem is that the wrong hands may just happen to be the technology itself.[42]

Second, the Asilomar AI Principles advocate self-healing with respect to the future risks of AI. Articles 21 and 22 address how to plan for future risks posed by AI. They state that AI can take the risks caused by its own development and can be controlled and improved through recursive self-improvement and self-replication. In short, these two principles oppose human intervention to control and adjust the development of AI through external intervention. An engineer-deterministic mindset

---

40 Ayoub and Payneb [41].

41 Bostolom [42].

42 Russell and Norvig [43].

pervades the development history of AI in the West. As John Markoff, a prominent American journalist, described, "Today, the engineers who design AI-based programs and robots will have a huge impact on the way we use them."[43]

Again, the Asilomar AI Principles do not provide detailed options and pathways for action, which will lead to a distortion of the principles in practice. Although Article 18 of the "The Asilomar AI 23 Principles" states that "An arms race in lethal autonomous weapons should be avoided", the principle does not provide a path to achieve this. The lack of a clear path to achieve this will lead to the principle being reduced to a hollow provision. In addition, Article 11 emphasizes "Human Values", that is, AI systems should be designed and operated so as to be compatible with ideals of human dignity, rights, freedoms, and cultural diversity. However, in practice, since the design of AI systems and the main discourse are in the hands of a few developed countries, the values embedded in them are likely to be the mainstream values of a few developed countries, rather than the values of human diversity as a whole. Similar issues arise in Articles 23, 14, and 15 principles as well. When it comes to the common good and shared prosperity, the problem of inconsistency between macro principles and actual practice arises.

In addition to the Asilomar Principles, Google released AI principles in 2018 that (i) Be socially beneficial; (ii) Avoid creating or reinforcing unfair bias; (iii) Be built and tested for safety; (iv) Be accountable to people; (v) Incorporate privacy design principles; (vi) Uphold high standards of scientific excellence; and (vii) Be made available for uses that accord with these principles.[44] The House of Lords Select Committee on Artificial Intelligence has published a landmark report on the future of AI in April 2018, which include: (i) Artificial intelligence should be developed for the common good and benefit of humanity; (ii) Artificial intelligence should operate on principles of intelligibility and fairness; (iii) Artificial intelligence should not be used to diminish the data rights or privacy of individuals, families or communities; (iv) All citizens should have the right to be educated to enable them to flourish mentally, emotionally and economically alongside artificial intelligence; and (v) The autonomous power to hurt, destroy or deceive human beings should never be vested in artificial intelligence.[45] The fifth of these five principles, while appearing to protect humans, actually affirms the autonomy of AI from another perspective. The EU released its AI ethics guidelines in 2019, including the following seven principles: Human agency and oversight; Technical robustness and safety; Privacy and Data governance; Transparency; Diversity, non-discrimination and fairness; Societal and environmental well-being; Accountability. Part of the progress of the EU AI ethical guidelines is their emphasis that "AI should not trample on human autonomy. People should not be manipulated or coerced by AI systems, and humans should be able to

---

[43] Markoff [44].

[44] Google Official website, *Artificial Intelligence at Google: Our Principles,* https://ai.google/principles, May 20, 2019.

[45] Verdict_AI, *The House of Lords AI Report: A Turning Point for British AI*, April 18,2018. https://verdict-ai.nridigital.com/verdict_ai_apr18/the_house_of_lords_ai_report_a_turning_point_for_british_ai, May 20, 2019.

Table 3.1 Comparison of important western ethical principles of artificial intelligence

Principles	Publishers	Release Time	Attitude towards the development of general AI	Attitudes toward AI autonomy	Emphasis on human autonomy
Asilomar 23 Principles	Future Life Institute	January 2017	Encourage the development of generic AI	Allowing AI autonomy	No explicit emphasis on human autonomy
Google AI Principles	Google	January 2018	No mention of generic AI	Allowing AI autonomy	No explicit emphasis on human autonomy
Five principles of AI code	The House of Lords Select Committee	April 2018	No mention of generic AI	Limiting the partial autonomy of AI	Emphasizing human autonomy by emphasizing citizens' right to education
AI Ethical Guidelines	EU	April 2019	No mention of generic AI	Acquiescence to AI autonomy	Explicit emphasis on human autonomy

intervene or oversee every decision that the software makes."[46] However, this is still a principle of passivity. In a way, this principle confirms the autonomy of AI systems (Table 3.1).

Second, the most influential ethical norms in the field of AI are currently defined by the West, which is the basis for policies, laws and related principles in reality. The most typical example of this is Isaac Asimov's "Three Laws of Robotics" (later referred to as "Asimov's Three Laws"), a representative of American science fiction. Although this law was proposed by Asimov in his science fiction novels, almost all of the aforementioned major Western principles can be understood as applicable or extended versions of this law. "Asimov's Three Laws" are: (1) A robot may not injure a human being or, through inaction, allow a human being to come to harm. (2) A robot must obey orders given by human beings except where such orders would conflict with the First Law. (3) A robot must protect its own existence as long as such protection does not conflict with the First or Second Law.[47]

The "Asimov's Three Laws" themselves are very contradictory. First, the first law emphasizes the subjectivity of robots. The first law emphasizes that "A robot may not injure a human being or, through inaction, allow a human being to come to harm," which means that robots are treated as unique legal or ethical subjects. However, the Second Law, which requires that "A robot must obey orders given by human beings," erases the subjectivity of robots. The Third Law, in turn, emphasizes that "A robot must protect its own existence" and reassigns the status of a robot subject. Therefore, Asimov's Three Laws are very vague and even contradictory in terms of subjectivity. The explanation for this contradiction can be found

[46] European Commission, *Artificial intelligence*, April 8, 2019. https://ec.europa.eu/commission/news/artificial-intelligence-2019-apr-08_en, May 20, 2019.
[47] Clarke [45].

in Western Christian culture. In Christian culture, God is "upper-case" and man is "lower-case". Only during the development of the modern Enlightenment was man gradually "upper-case", i.e., given subject status. In addition, Christian culture believes that man has original sin and that such a subjective status is unreasonable or unsustainable. In the Western world and logic, there is a metaphor that one day man's subjectivity will be taken away. Thus, the subjectivity of robots is readily accepted in the Western conception, and from the Western pessimistic worldview, many scholars readily accept the outcome of robots replacing humans. This is closely tied to the pessimistic Christian worldview. The Western scholarly mindset essentially embraces fully autonomous intelligences entirely or sees this trend as inevitable. Thus, in ethics, many Western scholars have advocated the realization of autonomous moral intelligences. For example, Colin Allen, a professor of ethics at Indiana University, and others argue that as artificial intelligence gets closer to the goal of a fully autonomous intelligence, the question of how to design and implement an artificial moral agent becomes increasingly urgent. Allen et al. also hope that such an artificial moral agent can pass the "moral Turing Test".[48] Of course, Western scholars have over-predicted the autonomy of the intelligence. And, who defines morality here? How to quantify morality and give it to machines to make decisions? These are all theoretical difficulties. It is difficult for human society to reach a consensus on a proposition like morality, so how to give it to a machine to make a decision?

When Westerners think about the relationship between man and machine, it is easy to fall into the master–slave dialectic. The current situation is that man is the master and the machine is the slave. But one day, the machine will strive to gain the status of master, and will even eventually rule the world, while man will become the slave of the machine. This is the natural application of Hegel's master–slave dialectic in the relationship between man and machine, and it can even be said to be the mainstream view in Western thought. Behind this view is the Western perception of the uncertainty of man's subject position. In the West, the "upper-case" human being did not appear until after modern times, and human subjectivity has only been highlighted in Western thought since modern times. Thus, the transient and fluid subjectivity forms the basis of the universal judgment of Western knowledge.

Asimov's First Law is that robots must not harm humans or sit back and watch humans get harmed. The underlying implication is that robots have their own subjectivity. If the subjectivity of robots is not recognized, then the expression of this law should be changed to "robots must not be used to harm humans". If robots are given too much decision-making power, or if subjectivity is given to robots, there is room for shifting responsibility or lack of responsibility. Specifically, if robots have some kind of subjectivity, then when robots harm humans, humans will emphasize that robots should be punished accordingly. However, robots themselves do not understand the meaning of punishment in the human world. At the same time, this subjectivity provides the possibility for other interested parties to shirk their responsibilities. For example, Marco Sassoli, a professor of international law at the University of Geneva, argues that one of the current difficulties with autonomous weapons is who exactly

[48] Allen et al. [46].

bears the responsibility for canceling an attack when it is found to be illegal in the case of its use.[49] At the same time, robots are run by programs and do not have their own emotions, which could easily develop into "killer robots" in a war. As the American scholar Peter Singer has pointed out, "Robots may invite a dark irony. They may seem to reduce human losses in war, but they may induce us to wage more wars."[50] In fact, the conflict of interests among human societies occurs mainly between people, and is a problem arising from the world of human society's own meaning, the responsibility and consequences of which should rightly be borne by human members, and not shifted to intelligences.

At the same time, this Western domination of ethical values is combined with its technological monopoly. For example, in the case of robotic weapons, developed countries want to develop technological superiority in robotic weapons and then maintain this superiority with the help of international norms. The typical attitude of the West is to promote the development of war robots first, and then regulate the development of war robots after technological superiority is developed, and freeze the technological superiority of developed countries by creating relevant international statutes to restrict the production and use of robots.

Thus, with respect to hypothesis 1-c, the current preliminary conclusion is that, as things stand, international rulemaking for AI is still completely lacking in developing countries, and there is no indication that these rules will move in the direction of favoring developing countries.

3.4 De-escalation or Aggravation? The Role of Artificial Intelligence on the Logic of Conflict

With respect to hypothesis 2-a, "The development of AI helps to increase the understanding between different groups", the relevant facts that can be observed so far are:

First, machine translation can further improve the effectiveness of dialogue among civilizations. The intercultural divide is a major cause of global problems, as discussed in Part I. Samuel Huntington also predicted that the clash of civilizations would become the main conflict of international society after the Cold War.[51] Cross-cultural communication has been a problem before the intelligent revolution. Due to the difference in language, people created barriers in communication, which in turn deepened the mistrust among people. Thus, cultural differences have been a major cause of ongoing conflicts and wars throughout human history. In the biblical story of the Tower of Babel, God dismantled the human alliance that built the Tower of Heaven by setting up a language barrier. People who do not speak the same language gradually form different cultures, and the cultural differences lead to the inability of

---

[49] Sassoli [47].

[50] Singer [48].

[51] Huntington [49].

people to understand each other, and this leads to conflicts and even wars. However, important advances in machine translation technology in artificial intelligence will greatly facilitate cross-cultural communication. It is well known that it is very difficult for people to learn new languages, and even more difficult to master multiple languages. However, the development of artificial intelligence has provided a chance for human to learn languages. On the one hand, machine translation can eliminate the difficulties in daily communication between people of different countries, and on the other hand, AI can also enhance human skills in turn, i.e. people can learn further and faster in the environment created by machines and improve their own language skills. For example, Chinese AI company LAIX can help people enhance their language skills through conversations with robots.

Second, intelligence-related technologies have a strong empowerment function for disadvantaged groups, which allows disadvantaged groups to open a more equal dialogue with other groups after empowerment. Disadvantaged groups often have certain deficiencies of their own, and AI technologies can precisely compensate or re-empower these deficiencies. For example, Radio Frequency Identification (RFID) technology can help blind people walk safely in certain environments.[52] Another example is 3D printing technology that facilitates the customization of prosthetics for people with physical disabilities.[53] Brain-computer interface technology gives the seriously ill paralyzed person the possibility to recover again. Amartya Sen, a renowned economist and philosopher, places great emphasis on equality of ability. According to him, true equality can only be achieved if there is equality of ability between people.[54] To achieve this equality of ability, artificial intelligence is undoubtedly the most important technological solution at present. In the past, it was difficult for people with impairments to integrate into public society, but technological advances are gradually turning this impossibility into a possibility.[55] There are many global issues such as health and poverty that involve the interests of disadvantaged people, and the development of artificial intelligence can provide new ideas for solving precisely these problems.

Third, after AI empowerment, the possibility and frequency of direct encounters between different cultural groups will increase, which may then cause new conflicts or adaptation problems in the short term. Although in the long term, direct dialogue due to technological empowerment helps different groups to communicate and vulnerable groups to express their interests, there will be a peak of expression in the short term. Difficulties and even conflicts in cross-cultural communication can also occur when different cultural groups meet directly. For example, research

---

[52] This technological system accomplishes navigation by providing voice guidance about the user's current location and navigation information about how to move to a specific location. A wireless sensor network (WSN) acts as the data transmission backbone by using RFID tags for localization to mark points. Tee et al. [50].

[53] 3D printing technology is one of the solutions for manufacturing hand prostheses and can be personalized for patients printing prostheses. Kate [51].

[54] Sen [52].

[55] According to statistics, in 2013 alone, 17.6% of people with disabilities in the United States successfully entered the workforce through assistive technologies (ATs). Field and Jette [53].

by Carole-Anne Sénit, a researcher in sustainable development and international relations at Utrecht University in the Netherlands, and others has shown that civil society is increasingly using information and communication technologies to engage in intergovernmental negotiations related to sustainable development, and that these technologies are also often seen as a potential democratic legitimacy deficit in traditional mechanisms remedy. However, in an empirical study of the extensive online dialogue at the 2012 UN Conference on Sustainable Development (Rio+20), Sennett et al. conclude that while ICT has contributed to dialogue and debate, it has reinforced, rather than reversed, implicit participatory inequalities globally and failed to significantly increase transparency and accountability. This, in turn, has prevented civil society from meaningfully participating in intergovernmental negotiations, thus demonstrating the limits of "cyber democracy".[56]

Regarding hypothesis 2-a, the current preliminary conclusion is that the development of AI increases the opportunities for communication between different groups and also allows different groups to express their wishes more directly, which in the long run is helpful for cultural exchange in global governance and for the elimination of differences, but in the short term, it may instead lead to a concentration of new types of conflicts because of the maladjustment brought about by direct encounters emerge.

With regard to hypothesis 2-b, "The development of AI contributes to the global free movement of populations", the relevant facts that can be observed are:

First, the development of artificial intelligence will likely exacerbate the global unemployment problem and make the anti-immigration wave even more powerful. Economists generally divide unemployment into two categories: cyclical unemployment and structural unemployment. Cyclical unemployment is a state where people cycle between employment and unemployment due to the external environment of economic recession and economic boom. Structural unemployment, on the other hand, describes the long-term, structural factors that eventually lead to the complete disappearance of certain occupations and jobs. The biggest impact of the intelligent revolution on employment is that it not only threatens physical workers in the traditional sense, but also has a clear substitution effect on mental workers. In the words of Jerry Kaplan, professor of artificial intelligence and ethics at Stanford University, "no matter what color your collar is, automation is relentless."[57] What human society faces in the future is likely to be global unemployment. Unemployment will be a problem not only for developing countries, but also for developed countries. And if new unemployed people move globally in the form of economic refugees, it will create even more global problems. There are already some anti-globalization or counter-globalization trends, such as the increasing tendency of developed countries to tighten their immigration policies in order to stop the influx of people from developing countries. In this light, the unemployment problem may resonate with the anti-immigration wave. The Hansen et al. study argues that machine-driven production of data, and the tools and techniques to interpret them, are created by highly specialized

---

[56] Sénit et al. [54].

[57] Kaplan [55].

individuals working in relatively opaque government or commercial organizations, and whose methods are not subject to traditional scientific scrutiny. As a result, the real risk of unemployment to society posed by AI is likely to go unnoticed by the general public for a longer period of time.[58]

Secondly, under the resonance effect, there will be regions in the world that strive to achieve closure. The free movement of people is the basis of global governance.[59] However, there is already a trend in the West against the free movement of people. Previously, developed Western countries encouraged the influx of people because of their labor shortage. Most countries have declined fertility rates after entering developed societies, leaving a shortage of subsequent labor, which then relies on the inflow of foreign labor. However, the biggest problem with the inflow of foreign labor is that it leads to the balkanization of society. It is such a concern expressed by Huntington in his book "Who We Are".[60] As a result, the Western world has been trying to reduce its dependence on foreign populations through the technological revolution. In this sense, the development of artificial intelligence technologies will curb those population movements that arise in search of work around the world. At the same time, one of the main points of the current Western far-right ideology is the opposition to population movements. They argue that immigrants from less developed countries are taking jobs away from the indigenous population and causing many social problems.[61] If the Western ideology against population mobility and immigration continues to develop, the current state of population mobility development worldwide will be frozen to some extent, and factor exchange between developed and developing countries will no longer occur. In this sense, the development of artificial intelligence and the anti-immigration wave will be untimely combined.

Regarding hypothesis 2-b, the preliminary conclusion so far is that the development of AI makes it easier to communicate when people are globally mobile, but at the same time there is likely to be a wave of anti-immigration everywhere under the increased unemployment problem, so this in turn is not conducive to the free global movement of people (see Fig. 3.2).

3.5 Global Good Intelligence: An Effort to Eliminate the Logic of Hegemonic

After synthesizing the conclusions of hypothesis 1-a, 1-b, and 1-c, the overall reality of hypothesis 1, "the development of AI helps to reduce the logic of hegemonic," is that there is no clear indication that the development of AI will lead to a decline in the dominant power of Western countries such as the United States. While developing

[58] Hansen and Porter [31, p. 32].

[59] Qiqi [56].

[60] Huntington [57].

[61] One of the reasons for the UK's departure from the EU was the fear of a series of problems caused by too many new immigrants. Abrams and Travaglino [58].

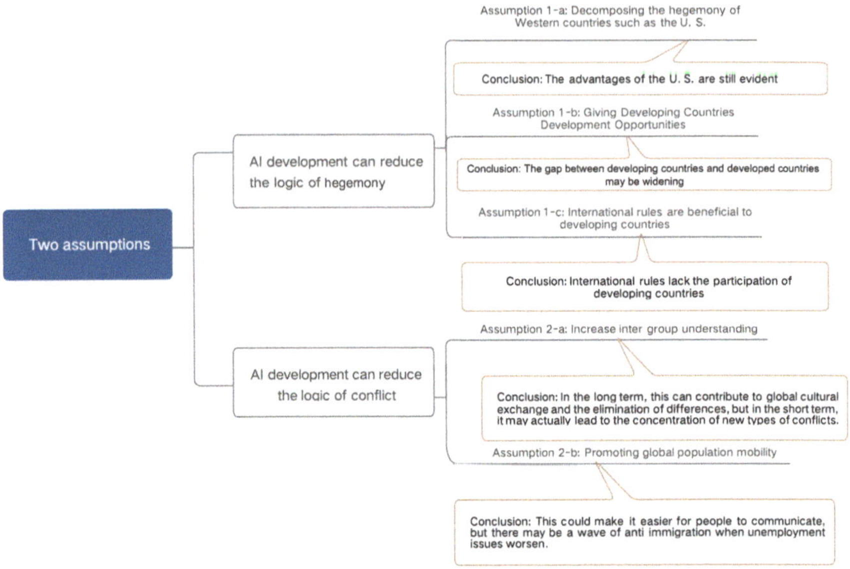

Fig. 3.2 Can the development of artificial intelligence change the two Crux logic of global governance

countries can take second-mover advantage, they face many of the same dilemmas and may encounter technological blockage from the West. At the same time, there is no evidence that the international rules of AI are moving in the direction of favoring developing countries. This conclusion tells us that the application of AI in global governance does not naturally lead to the dissolution of the logic of hegemonic. In a sense, the logic of hegemonic has even been somewhat reshaped and strengthened. Therefore, we need to make efforts at the value level to reach a new consensus in the international community to jointly eliminate the logic of hegemonic.

Based on this, the author believes that the targeted value of global smart governance should be global good intelligence. Global good intelligence is related to and different from global good governance. Global good governance is an important theory in global governance theory, which refers to achieving the goal of global justice through good governance. According to Professor Yu Keping, "Good governance is the process of social management that maximizes the public good."[62] In the age of intelligence, we likewise need to further consider such value-based issues. For example, what is the original purpose of our development of artificial intelligence? In this sense, good intelligence is the intelligence of goodness, i.e., defining the goal of AI development as goodness and justice. And global good intelligence is about spreading the goal of justice in AI at the global level and further expanding the positive outcomes of global governance. Luciano Floridi, Professor of Ethics at the University of Oxford, proposes the concept of "ethical globalization": "Ethical discourse

[62] Keping [59].

suggests that it needs to be updated to respond to a globalized world in which its parts are closely interconnected. Every ethical theory is required to defend its global and cross-cultural relevance."[63] This new requirement of ethical globalization in the age of intelligence can then be summarized as global good intelligence.

Further, global good intelligence should contain the following two connotations. On the one hand, the development of AI should further promote equality and justice in the global society. Previously, John Rawls emphasized distributive justice at the domestic level in "A Theory of Justice", while Thomas Pogge, a professor of philosophy at Yale University, and Charles Beitz, a professor of political theory at Princeton University, have raised the issue of international distributive justice on this basis. Beitz argues, "If the difference principle is chosen out in the domestic original position, similarly the parties should choose the difference principle in the international original position."[64] That is, nationals of countries also have the obligation to balance distribution among themselves. Berger also endorses the application of Rawls' contractarianism and distributive justice to the global society, and uses the situation of the most disadvantaged groups globally as a reference for institutional judgment.[65] How to define global justice in the age of intelligence and how to further promote good intelligence in the age of artificial intelligence are questions that deserve deeper consideration. On the other hand, the purpose of good intelligence is to develop benign AI, implying that the development of AI should promote the solution of global problems rather than exacerbate them. As Wiener cautions, "The new industrial revolution is a double-edged sword that can be used for the benefit of humanity, but only if humanity survives long enough for us to enter this period of benefit to humanity. The new industrial revolution can also destroy humanity, and if we don't use it wisely, it has the potential to grow to that point very quickly."[66] Kareem Ayoub and Kenneth Payneb, researchers at the artificial intelligence firm Deepmind, argue that AI is facing the kind of ethical dilemma that human intelligence currently faces.[67] Based on this, how to define good intelligence and global good intelligence is particularly critical.

Specifically, achieving global good intelligence requires efforts in the following areas:

First, the global community needs to reach a consensus on the development of general AI. "The Asilomar AI principles" emphasizes the feasibility of developing general AI, which is also the consensus of the Western business community. They hope to solve all problems once and for all through the development of general AI.[68] However, the development of general AI may eventually lead to challenges to the meaning of human existence. Therefore, the global community should form a consensus on the development of general AI, i.e., to conclude which types of general

[63] Floridi [60].

[64] Beitz [61].

[65] Pogge [62, 63].

[66] Wiener [64].

[67] Ayoub and Payneb [41, p. 811].

[68] Müller and Bostrom [65].

AI can be developed and which ones cannot. The author believes that, on the one hand, we cannot completely suspend the development of general AI, because general AI can indeed help humans solve certain problems more efficiently and effectively in many ways. Moreover, since the development of AI is basically a competition among general AI at a later stage, and all countries will make efforts in this field, it is likely to lose the competitive advantage by rashly giving up the development of this field. On the other hand, it is important to fully assess the types of general-purpose AI and the overall development consequences. In particular, it is important to be sufficiently vigilant about the development of higher levels of general AI, i.e., super AI, because such an outcome is likely to challenge the world of human meaning.

Western scholars are almost always completely pessimistic when describing the political consequences of AI. Ayoba and Penib argued that in the long run (estimates among professionals range from 20 to several hundred years), AI can develop general intelligence capabilities that match or even far exceed those of humans, leading to complex subjective values measured on different issues.[69] For example, Ivana Damnjanović of the School of Political Science and International Relations at the Australian National University argues that political theory always seems to lag behind technological developments. With the field of artificial intelligence rapidly evolving, a common estimate is that the technological singularity is likely to occur within the next 50 to 200 years. Even without considering the time frame, the possibility of super-human intelligent AI raises serious political questions. Damnjanović discusses the science fiction novel *Polity* by British science fiction author Neal Asher. In Asher's welfare society, the fate of humanity is ruled by benevolent artificial intelligence, and politics becomes a thing of the past. According to Damnjanović, this seems to be a rule that is in the best interest of all. However, such a world is defined not by human rules, but by machine rules.[70] The Hansen team's research argues that the growing dominance of these self-subjects will cause them to gradually replace human cognitive abilities. Although these machine-driven processes are ultimately made by humans, these processes can make human agency ultimately secondary.[71] For scholars such as Ayouba, future development goals for AI also include the generation of increasingly complex intelligence through successor systems, which may require additional resources. Thus, there will be more serious tensions between AI and humans in terms of competing resources to meet these goals, or in terms of the ethical guidelines to follow in reconciling these tensions.[72]

Based on this, we believe that a "new three principles of robotics" (later referred to as "new three principles") should be proposed: First, robots will always be auxiliary; second, the proportion of human decision-making should not be lower than the golden ratio; third, humans should always grasp the rhythm of artificial intelligence development and be ready to pause or slow down. Under the regulation of the "New Three Principles," the existing controversies about robots can be resolved to

---

[69] Ayoub and Payneb [41, p. 816].

[70] Damnjanović [66].

[71] Hansen and Porter [31, p. 33].

[72] Ayoub and Payneb [41, pp. 813–814].

some extent. For example, according to the first rule of the New Three Principles, artificial intelligence can be given a legal status, but not a full, human-like status. The granting of citizenship to the robot Sophia by the Saudis can be seen as an important landmark event. In a similar vein, driverless systems were given the status of driver in a groundbreaking push to test driverless cars in the United States. The issue of electronic human rights, which has been discussed in Europe in recent years, emphasizes the need to grant AI systems some kind of human-like rights relationship and legal status. This is in spite of the fact that some scholars advocate for the proposal to grant full human status to intelligences. For example, Martine Rothblatt, an American lawyer, author, and entrepreneur, asserts that "we are best prepared for tomorrow's world when we achieve respect for others (i.e., virtual people) as we do for ourselves and spread this virtue everywhere in the world."[73] The author believes that this should not be the case. An intelligent being should be a "semi-sovereign person", i.e., an intelligent being can be given a certain identity. Here, we can refer to the way companies are recognized. In corporate law, a corporation is given the status of a legal person because it is a fictitious person. In this sense, an intelligent body can be given a certain status, but at the same time, this status cannot be exactly the same as the citizenship of a human being.

According to the second principle of the "New Three Principles", AI can be an aid to human decision-making, but humans cannot leave all the decision-making in society to AI. The third principle of the "new three principles" will guide people to be more cautious and optimistic about the development of artificial intelligence, and always master the pace of development of artificial intelligence. The main purpose of the proposed human decision-making ratio of no less than the golden ratio is to ensure that humans dominate in future human–machine interaction, as well as robots in the decision-making support position. Judging from the current trend, humans are increasingly accustomed to leaving almost all their decisions to machines. Taking map navigation as an example, although the development of navigation systems can greatly facilitate people's travel, the act of relying entirely on navigation will likely lead to human degradation. Therefore, it is important to improve the efficiency of human decision making by using machines, rather than leaving it entirely to machines. In this sense, by setting the golden ratio, we can be repeatedly reminded that we cannot leave all decisions to machines. The golden ratio of 0.618 is chosen because when machines are once more than half in decision making, their existence will probably threaten the long-term development of human beings and the meaning of human beings. So the existence of the golden ratio can provide an important reminder function to human society. Of course, in practice, it is a very complicated problem to evaluate the ratio of human decision making to machine decision making. This problem involves a large number of details and the calculation is very complicated, so limited by the space of this paper, the author will study it in depth in the future research.

Western scholars hope to discipline machines by achieving their autonomous morality. Ayouba and Penibu point out that while AI may have emotional and moral

[73] Rothblatt [67].

capabilities, it may also not. Faced with this possibility, humans would do well to complete their strategic planning first.[74] Western scholars generally define this strategic planning as the development of ethical machines. For example, Wendell Wallach and Colin Allen, advocate making machines capable of knowing right from wrong by building artificial moral intelligences, i.e., moral machines.[75] However, the author argues that this mainstream thinking among Western scholars may have been wrong from the beginning. Since moral machines are to be cultivated, the premise is to recognize the subjectivity of machines without reservation. And, the debate about the connotation of morality has not ceased throughout the history of philosophy. So whose morality belongs to the moral machine? What kind of morality is it? The result we do not want is that morality here becomes morality as defined by mainstream Western values.

The reason why the subsidiarity of AI is emphasized here is actually to further highlight the subjectivity of human beings. Human subjectivity needs to be confirmed and embodied in practical activities. Through productive labor, man changes the form of existence of nature and realizes human conceptions and goals, thus putting nature under his own subjective control and becoming the master of nature. In this regard, Marx states, "so in my individual activity I would have directly confirmed and realized my authentic nature, my human, communal nature."[76] Centrality here is the process of creative practice. In the midst of this process, the essence of man is intuitively present: "It is just in his work upon the objective world, therefore, that man really proves himself to be a species-being."[77] The problem with AI alienation, however, is that the subjectivity and practicality of man is fundamentally challenged. In the Economic and Philosophical Manuscripts of 1844, Marx pointed out that "free, conscious activity is man's species-character."[78] This "free, conscious activity" is the essence of man. If this characteristic is lost, man will hardly be human.

Second, the development of intelligence should be paced through a global consultation mechanism. The development of AI is currently driven primarily by national states or their internal firms, yet the impact of AI is global in nature. For example, competition among countries over AI development can create a global-level competitive anxiety that it is difficult for a country to slow down in this area because slowing down may mean being out of the game. This competitive anxiety may invalidate the third principle of the "New Three Principles of Robotics" proposed earlier. Therefore, it is crucial to develop a global level of institutional coordination. We cannot coordinate globally only after a clear threat has emerged. During the Cold War, the U.S. and Soviet Union coordinated only when the threat was clear and they felt that

[74] Ayoub and Payneb [68].

[75] Wallach and Allen [69].

[76] *Karl Marx, Frederick Engels: Complete Works*, Vol. 42, Beijing: People's Publishing House, 1979, p. 37. (in Chinese).

[77] *Karl Marx, Frederick Engels: Complete Works*, Vol. 3, Beijing: People's Publishing House, 2002, p. 274. (in Chinese).

[78] *Karl Marx, Frederick Engels: Complete Works*, Vol. 3, Beijing: People's Publishing House, 2002, p. 273. (in Chinese).

the weight of destruction could no longer be borne. Both the U.S. and the Soviet Union initially wanted more nuclear weapons, but once both sides realized that the total number of nuclear weapons would be enough to destroy the entire human civilization thousands of times over, they began to negotiate a nuclear arms race in order to jointly limit the size of nuclear weapons.

Then again, the unemployment problem likewise requires a globally coordinated layout in advance. Appropriately preserving some redundant settings in human society or taxing AI systems and compensating the unemployed can buy more time for human society's own regulation to slow down the rate of replacement of human occupations by intelligences.[79] Otherwise, human society is likely to be caught in a dilemma of self-fighting or even intensified conflicts. As mentioned earlier, if there is a lack of moderate regulation of AI development, then the North–South gap may further widen and eventually lead to further intensification of global problems. Unemployment can likewise extend beyond the nation-state, and thus the significance of global coordination becomes even more pronounced.

Third, interpretable and secure artificial intelligence will be the future direction. As Marx pointed out, "Natural science will in time incorporate into itself the science of man, just as the science of man will incorporate into itself natural science: there will be one science."[80] The development of artificial intelligence involves not only the natural sciences but also the social sciences. Artificial intelligence is not only the science of robots, but also the science of people, because artificial intelligence has to be applied to human society. Because of this, IBM Watson cognitive computing platform consultant Amir Husain believes that to advance the development of artificial intelligence, it is urgent to set ultra-high security and interpretability standards.[81] In short, the future development of AI has to move closer to the human characteristics of safety, humanity and rationality. The main achievement of this wave of AI is attributed to deep learning algorithms in machine learning. Deep learning is data-based, and feature quantities are automatically generated by computers. It does not require a human to design the feature quantity, but rather the computer automatically acquires the high-level feature quantity.[82] A typical deep learning model is a deep neural network. For a neural network model, a simple way to increase the capacity is to increase the number of hidden layers. The number of hidden layers increases, then the corresponding neuron connection rights, thresholds and other parameters will be increased, and the complexity of the model can be increased.[83] Deep learning emphasizes the abstraction of data, automatic learning of features, and the importance of connectionism. However, deep learning lacks a sound theory. Therefore, in the application practice of deep learning, engineers need to manually tune the parameters to get a good model, but at the same time, these engineers cannot explain

[79] Qiqi [70].

[80] *Karl Marx, Frederick Engels: Complete Works*, Vol. 3, Beijing: People's Publishing House, 2002, p. 308. (in Chinese).

[81] Hussain [71].

[82] Matsuo [72].

[83] Zhihua [73].

the factors affecting the model effect.[84] In other words, due to the uninterpretability of a large number of internal parameters, there is an algorithmic black box in itself. Therefore, the development of interpretable AI is the key. Different from the path of deep learning, knowledge graphs hope to make a breakthrough in the development of explainable AI by means of entity linking, relation extraction, knowledge inference and knowledge representation.

Fourth, we promote the solution of historical problems in developing countries through intelligence. Good intelligence is more about empowerment than replacement. Through the tremendous empowering power of AI, disadvantaged people in developing countries can have more opportunities to improve their conditions. Kaplan believes that American society needs to think deeply about distribution if it is to cope with the impact of the intelligent revolution. Kaplan notes, "We don't need to take anything away from anyone, we just need to distribute future growth in a more equitable way and the problem will be solved."[85] This view of Kaplan is insightful, and at the same time we need to extend and apply this view in the international community. In this sense, intelligence should be the optimal solution to most global problems. For example, many developing countries suffer from water scarcity, which can be effectively solved by effective and intelligent water management solutions. For example, for developing countries, the biggest problem is the influence of traditional factors on management activities. Intelligent solutions can minimize the impact of traditional factors with the help of intelligent systems and equipment, thus effectively addressing some of the historical problems of developing countries. Another example is that intelligent technological means can make government regulation more efficient and transparent, which can reduce the possibility of corruption while controlling administrative costs. Therefore, artificial intelligence should be promoted as an important solution and idea to solve global problems. The biggest advantage of artificial intelligence is that it can greatly save human resources, and the biggest difficulty for the United Nations and other international organizations to carry out relevant tasks in third world countries is the lack of human resources, so artificial intelligence can play an important supplementary role in this regard.

Fifth, we should strike a balance between global and national governance.[86] AI-related global mechanisms need to be formed at the global level and interact positively and benignant with domestic systems. On the one hand, at present, international rules of AI are mainly promoted by enterprises or institutions in developed Western countries. This means that a global AI governance mechanism considered from the global level as a whole is missing, so the next step requires developed and developing countries to jointly promote the construction of such a mechanism. On the other hand, global governance ultimately has to return to the framework of national governance. In the global society, the state remains the most important actor. Most of the key issues related to individual welfare and safety and security ultimately

[84] Zhiyuan [74].

[85] Kaplan [75].

[86] On the balance and interaction between global governance and national governance, see Tuo [76]; Zhimin [77]; Xuelian and Lu [78]; Zhicheng [79]; Zhenye [80].

have to be solved by national subjects.[87] Therefore, a program that talks only about global governance without national governance is hollow and impractical. Research by Indiana University law professor Fred H. Cate points out that government agencies have shown a strong interest in the use of big data in areas such as identifying financial transactions related to money laundering and terrorism, locating criminal suspects, and identifying and blocking child pornography. This implies that we need to rethink the concept and connotation of national sovereignty in the context of global data flows.[88] At the level of national governance, the most critical question is how to improve the national governance capacity of countries (especially developing countries) through the technological and industrial development of artificial intelligence, so as to fundamentally eliminate the domestic symptoms that cause global problems.

3.6 Global Collective Intelligence: An Effort to Eliminate the Logic of Conflict

After combining the findings of hypothesis 2-a and 2-b, the basic reality of hypothesis 2, "the development of AI facilitates the reduction of conflict logic," is that in the long run, the development of artificial intelligence will help people to understand more, but in the short run, direct and frequent contact may increase new conflicts. At the same time, AI technologies may help people on the move to understand other cultures, but the anti-immigrant wave exacerbated by global unemployment is not conducive to the free global movement of people. This conclusion suggests that the use of AI in global governance can hardly lead directly to a decline in the logic of conflict either, or even exacerbate it in the short term. Therefore, preparation at the value level is very important. The key question here is how to form a tight and united global community in the age of intelligence.

In this process, China's role becomes particularly crucial. China has long emphasized itself as a member of the Third World.[89] Although China has used the expression "third world" less frequently in international contexts since its reform and opening up, it still emphasizes itself as a representative of the developing world. In the report of the 19th CPC National Congress, despite declaring a new era, China still emphasizes that it is a member of the developing world and is still in the primary stage of socialism.[90] In fact, China is also actively fighting for the interests of the developing

---

[87] Qiqi [81].

[88] Cate et al. [82].

[89] In a conversation with Zambian President Kaunda on February 22, 1974, Mao Zedong put forward a theory about the division of the three worlds. Mao said, "I see the United States, the Soviet Union as the first world, and the centrists, Japan, Europe, Australia and Canada, as the second world. We are the third world." "The Second World, Europe, Japan, Australia, Canada", "Asia is the Third World except for Japan. The whole of Africa is the Third World, and Latin America is also the Third World." See *Selected Writings of Mao Zedong*, Vol. 8, Beijing: People's Publishing House, 1999, pp. 441–442 (in Chinese).

[90] Jinping [83].

countries in world politics. For example, China has proposed the "The Belt and Road" to share its development experience with other developing countries, and to provide assistance to other countries in infrastructure development and livelihood services on the basis of the "Five-Pronged Approach".[91] From these measures, we can see that China is considering its own interests as well as the interests of the developing countries in the international political arena.

On the issue of AI development, China's attitude contrasts with that of the West in two ways: First, unlike the pessimism of the West, China is cautiously optimistic about the development of AI. Both academia and industry in the West are very pessimistic about the future outcome of AI. Kaplan, for example, describes such a picture of the future: "The earth may become a zoo without walls, a solid breeding ground for terrestrial animals, where there is only sunlight and solitude, where our mechanical caretakers will occasionally push us a little in order to maintain normal operation, and where we will welcome such help with our hands held high for our own well-being."[92] This point is closely related to the eschatology of the Western Christian cultural tradition.[93] This pessimistic view is not only reflected in the treatment of the development of artificial intelligence technologies, but also appears in the ethical discussions of other cutting-edge technologies, such as biotechnology. For example, the Japanese-American scholar Francis Fukuyama argues that, driven by biotechnology, humanity may enter a post-human future that may be more hierarchical and contradictory.[94] China, by contrast, holds more of a cautiously optimistic view. For example, China's emphasis on the need to seize the opportunities of the new technological revolution and proactively develop related science and technology and their applications reflects China's optimism in the development of AI. At the same time, China's emphasis on actively addressing the ethical and legal implications of AI is a cautious approach to the development of AI. As President Xi Jinping pointed out, "Ensure that AI is safe, reliable, and controllable. We should integrate multidisciplinary forces, strengthen research on legal, ethical, and social issues related to AI, and establish sound laws and regulations, institutional systems, and ethics that guarantee the healthy development of AI."[95]

Second, unlike the West's emphasis on exclusivity and hegemony, China places more emphasis on sharing with other countries (especially developing countries). If the development of AI eventually leads to the end of the world division of labor system, then the possibility for developing countries to move toward development through exchanges and cooperation with the Western world may be ruined. If the West further restricts the inflow of third world populations, then the flow of AI technologies to the third world becomes even less likely to materialize. China, however, has a clear attitude of extensive consultation, joint contribution and shared benefits. "The Belt

[91] Wen et al. [84].

[92] Kaplan [85].

[93] Qiqi [86].

[94] Fukuyama [87].

[95] Jinping [88]

and Road" initiative is representative of this philosophy.[96] President Xi Jinping states, "'The Belt and Road' construction upholds the principle of extensive consultation, joint contribution and shared benefits, and is not closed, but open and inclusive; it is not a solo performance by China alone, but a chorus of countries along the route."[97] China is actively initiating cooperation, synergy and sharing of frontier technologies in "the Belt and Road" construction. Based on the attitude of extensive consultation, joint contribution and shared benefits, the development of artificial intelligence can benefit and facilitate the vast majority of the global community. As a representative of developing countries, China is able to assert the interests and aspirations of developing countries in the era of AI. Therefore, China's development of AI has a moral stance, i.e., China develops AI not only for its own sake, but also for the development rights and interests of the majority of developing countries.

Based on this, I propose the concept of global collective intelligence. Global collective intelligence is the collection of the wisdom of human beings and intelligences on a global scale to contribute to the solution of global problems. It is a kind of holistic thinking, i.e., considering problems from a global perspective. At the same time, global collective intelligence is also a kind of parallel thinking. The main reasons for the major breakthroughs in AI are the increase in the amount of data, the increase in computing power and the improvement of algorithms. In fact, machine algorithms are very clumsy compared to humans, but because machines have powerful parallel computing capabilities, it is possible to integrate relatively clumsy algorithmic steps into a holistic and excellent result at the end. If humans also learn this kind of parallel computing from machines, unite the strengths of all parties, reach a consensus, and work together, they may be able to overcome the global problems that plague people. As President Xi Jinping pointed out, "We should seek wisdom and draw nourishment from different civilizations, provide spiritual support and solace for people, and join hands to solve the various challenges faced by humanity together."[98] In this sense, it is necessary to form a "new global governance consensus" in the era of artificial intelligence.[99] Specifically, the global collective intelligence mainly contains several aspects:

First, human–machine collective intelligence. This has two main elements: first, the respective advantages of artificial intelligence and human intelligence can be complementary. The advantage of artificial intelligence is that it can be replicated, be easy to promote, make information storage and memory recall more complete, but also effectively avoid the interference of irrational factors such as emotions. Of course, artificial intelligence also has some shortcomings. As a technology-intensive and capital-intensive industry, AI needs to balance the costs and benefits of its application and development. In addition, current deep learning algorithms are uninterpretable due to the existence of an explanatory black box, which leads to uninterpretable results. In contrast, human intelligence, which is inherently general-purpose

---

[96] Kang [89].

[97] Jinping [90].

[98] Jinping [91].

[99] Bin [92].

intelligence, will be more efficient in facing more contexts and dealing with more complex problems. However, human intelligence requires alternating between rest and work to achieve it, and is also unable to achieve large scale expansion in a short period of time, and suffers from fuzzy memory as well as perceptual factors driving it. Therefore, in practice, human intelligence and artificial intelligence can complement each other and take advantage of each other's strengths. Second, AI can be used as an assistant, friend, and supporter of humans, but important decisions should still be made by humans. The machine should always be a supporter, not a decision maker. In some fields, machine decisions can be made to become dominant, but the overall ratio of machine decisions should not exceed the golden ratio, otherwise the world of meaning will be blurred and eventually lead to the disappearance of human uniqueness.

Second, multinational collective intelligence. In the current global governance, the state still has an important role. The main interests of individuals still need to be safeguarded by the state, and any mechanism of global governance will ultimately have to be implemented at the national level. Judging from the relationship between the previous industrial revolutions and the state, if the competition for technological dominance is not handled properly, then the final result is likely to be a war between states. This can be seen in the First and Second World War. Signs of conflict between the United States and the Soviet Union, as well as between the United States and Japan, can also be seen in the Third Industrial Revolution. In the development of the Fourth Industrial Revolution, the United States is the traditional rule-definer, while China is the emerging force. As a result, the United States has come to view China as an important competitor to contain, and this has been extremely evident in the U.S.-China trade dispute in recent years. The essence of the U.S.-China trade dispute is technological competition. The U.S. is using the U.S.-China trade dispute in hopes of preventing China from gain dominant or advantageous position in the new technological revolution.[100] Obviously, the U.S. still uses the traditional logic of Cold War thinking to look at the current problems, and the ultimate orientation of this Cold War thinking is bound to be conflict, which can be reflected in the attitude of Britain towards Germany before World War I. We precisely have to jump out of this wrong thinking. Only by breaking through the Cold War mentality and combining the strengths of China and the United States under a new framework of "collective intelligence" can we jointly contribute to the scientific and technological progress and development of mankind.

This Cold War mentality in the era of Big Data is reflected in data nationalism. Daniel Castro, vice chairman of the Information Technology and Innovation Foundation, a US think tank, believes that in the view of data nationalists, data can only be safe if stored in their own country. This perception is increasingly being absorbed by national policy makers and integrated into national data-related laws and policies. Castro argues that this misguided view will eventually lead to a series of policies

[100] Shipping [93].

that impede innovation, productivity and trade.[101] In response to this misguided view, multinational collective intelligence would require countries to form a series of conventions on the application of AI in global governance. For example, U.S. scholar John Weaver suggests that countries should form a series of international conventions on AI and state responsibility, national sovereignty, self-navigating vehicles, intellectual property rights, surveillance, and armed conflict. Weaver notes that multilateral agreements should be drafted among nations to determine how AI affects national sovereignty, what level of AI drone surveillance is permissible, and what AI is allowed to do in armed conflict.[102] This multinational collective intelligence emphasis should not only be based on international conventions with fixed rules and binding commitments, but also provide more flexible soft governance frameworks, such as relying on voluntary "non-adversarial and non-punitive" compliance mechanisms.[103] ASEAN's approach to regional governance, for example, is known for its inclusiveness, informality, pragmatism, facilitation, consensus building and non-confrontational negotiations,[104] in contrast to the "hostile posture and legitimate decision-making processes of Western multilateral negotiations". The same is true of the Paris Agreement on climate change, which represents a new form of global governance. Thus, future multinational cooperation on AI could take place more in the form of soft governance.

Third, multi-actor collective intelligence. The future of AI-enabled global governance should not be determined by just a few countries or a few mega-corporations. A broader range of multi-actors should be included in the production and formulation of AI rules, policies, and laws. Overall, AI has a strong empowering effect on non-state actors. First, the intelligent revolution has made coordinated governance by international organizations less costly. For example, policy coordination and communication among states will be easier with the aid of technologies such as AI and blockchain, while the monitoring and inspection of state compliance with international organization commitments can likewise be further institutionalized through smart technologies. Floridi proposed the concept of a political multi-intelligence system, arguing that such a system could reduce resource waste and maximize returns, and that international organizations are more in line with the characteristics of a political multi-intelligence system.[105] However, because resources are concentrated more toward states and multinational corporations, the effectiveness of international organizations depends on the extent to which states delegate authority to them. Already, the first signs of a decline in globalization and the rise of statism are emerging, so the attitude of major Western countries will be crucial. Philip Alston, a professor of law at New York University, argues that countries at the forefront of developing lethal robotic weapons technology may be less willing to accept international law

[101] Daniel Castro, "The False Promise of Data Nationalism," December 9, 2013. https://itif.org/publications/2013/12/09/false-promise-data-nationalism, May 20,2019.

[102] Weaver [94].

[103] Acharya [95].

[104] Acharya [96].

[105] Floridi [97].

jurisdiction or to initiate legal or moral constraints. At the same time, the target or victim states of such technologies are unlikely to lead the effort. Inevitably, therefore, this responsibility falls to international actors. Alston believes that the United Nations should play an important role. The UN Secretary General should convene a group of military and civilian representatives composed of key national departmental officials in human rights and humanitarian law, applied philosophers and ethicists, scientists and developers, with the aim of promoting the goals associated with the development of these new technologies. The group is tasked with considering what approaches can be taken to ensure that these technologies comply with applicable human rights and humanitarian law requirements. This would include consideration of the principle that any drone or robotic weapons system should have the same or better safety standards than comparable manned systems, detailed instructions for testing the reliability and performance of such technologies prior to deployment, and the incorporation of recording systems and other technologies into developed weapons systems to allow for effective investigation and accountability for alleged improper use of force.[106]

Otherwise, individuals and social organizations can be empowered to some extent by the intelligent revolution, but they also face many challenges. For example, using self-media, individuals can also become the center of public opinion, thus influencing and countering the traditional mainstream media, which is in the hands of the state. Most of the products of the intelligent revolution are centered on the individual, the consumer, so this consumer empowerment is actually individual empowerment. Using new technological tools, individuals can do greater association at the global level. Dan Plesch of the University of London and Thomas Weiss, a leading scholar of global governance, see the global expansion of networks and informal institutions as a serious challenge to post-war multilateralism, particularly the mechanisms of the United Nations. They warn that enthusiasm for formal and systematic multilateralism should be maintained, rather than focusing too much on ad hoc and informal multilateralism.[107] Amitav Acharya, a professor of international relations at American University, criticizes Plesch and Wise's view that global governance today must include not only the major powers but also international and regional institutions, non-state organizations, cooperative organizations, and people's movements and networks.[108] In this sense, the Internet, big data and artificial intelligence have certainly contributed to the development of these informal networks. Frank Biermann, Professor of Global Governance for Sustainable Development at Utrecht University, and others argue that there is a trend toward fragmentation of global governance. This fragmentation consists of international institutions (organizations, institutions, and implicit norms) of different natures, their spatial scope (from bilateral to global), and their subject matter (from specific policy areas to universal concerns).[109] While Billman and others explain fragmentation more in its

[106] Alston [91].

[107] Plesch and Weiss [92].

[108] Acharya [93].

[109] Biermann et al. [94].

negative sense, other scholars argue that fragmentation represents precisely a new direction. For example, Stewart Patrick, director of the International Institutions and Global Governance Program at the Council on Foreign Relations, argues that this fragmentation has produced "Good Enough Global Governance".[110] Acharya agrees that this fragmentation is inevitable, even creative, because it reflects the broader forces of political change in the world. In more detail, today's world is culturally and politically diverse, but more interconnected and interdependent.[111]

In his latest article, Thomas G. Weiss seems to have adjusted his position. His article, in collaboration with Rorden Wilkinson, advocates an Everyday Global Governance, which attempts to think about global governance from the perspective of the governed, so that the complexity, time, space, continuity, and change of global governance can be understood more deeply.[112] In the process of global mechanism formation, the participation of individuals as global citizens in the decision-making process is also of great significance. Oliver Fox and Peter Stoett argue that citizen participation is critical to the effectiveness and legitimacy of global governance. Through an assessment of the roles of governments, civil society, business, knowledge-based institutions, and citizens in the formation of the UN 2030 Agenda, Fox and Stoett argue that citizen participation offers a new option for democratic global governance.[113] As Professor Ortwin Renn, Director of the Institute for Advanced Sustainability Studies in Potsdam, Germany, points out, modern societies need the participation of society at large more than ever before. Effective and adequate mechanisms for citizen participation can reduce protests and resistance to the future risks of new technologies (see Fig. 3.3).[114]

3.7 Concluding Remarks

In "Machines, Natural Force and the Application of Science," Marx clearly expresses the idea of science and technology for economic and social change: "In acquiring new productive forces men change their mode of production; and in changing their mode of production, in changing the way of earning their living, they change all their social relations."[115] Artificial intelligence, as a new productive force, is changing the way we produce. At the same time, people's social relations and governance structures are adjusting along with it. Therefore, the occurrence of the intelligent revolution requires us to rethink the concept of future global governance in a holistic manner. Global Good Intelligence can be the target value of future AI applications

---

110 Patrick [95].

111 Acharya [88].

112 Weiss and Wilkinson [96].

113 Fox and Stoett [97].

114 Renn [97].

115 *Karl Marx, Frederick Engels: Collected Works*, Vol. 1, Beijing: People's Publishing House, 2012, p. 222. (in Chinese).

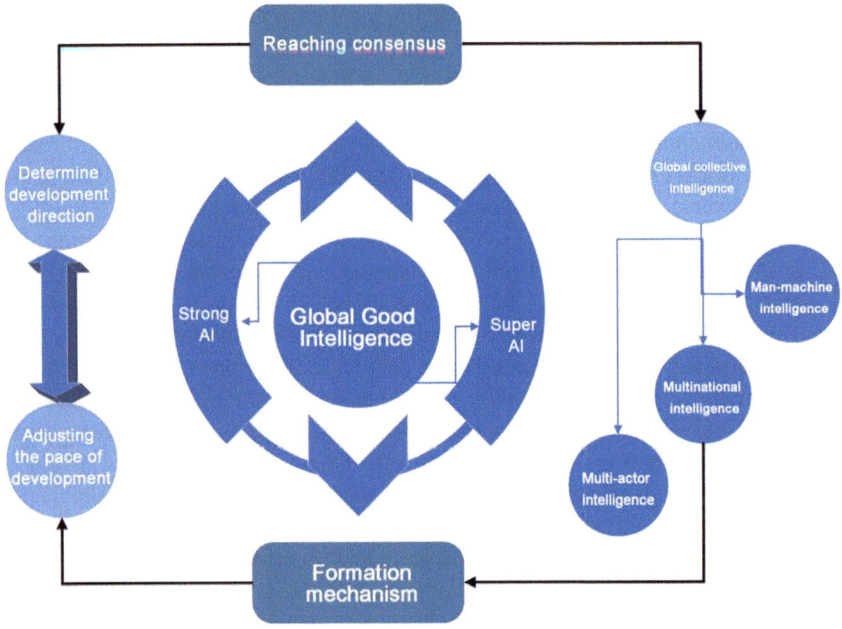

Fig. 3.3 Relationship between global good intelligence and global collective intelligence

in global governance. Global Good Intelligence has to address the question of where AI is going. Artificial intelligence is considered to be a disruptive technology. On the one hand, this technology may be useful for structural problem solving. On the other hand, because of its huge disruptive effects, the consequences of poor management can cause the capsizing of global society. At the heart of this is the issue of general AI. Because general AI ultimately challenges what it means to be human, it must be rationally and strictly restrained. It is for this reason that I propose the "New Three Principles of Robotics". Just as the rapid development of nuclear weapons during the Cold War caused panic throughout human society, a global consensus on the control of nuclear weapons was formed only afterwards. At present, the development of artificial intelligence is still in the early stage, and many disruptive effects have not yet been fully demonstrated. Therefore, research on the global governance effects of AI should be forward-looking in order to ensure the healthy development of AI and its benign application in global governance. The application of AI in global governance should emphasize not only the principle of controllability and regulation, but also the principle of differentiation, i.e., analyzing the problem more from the perspective of developing countries and vulnerable groups in global society.

Global collective intelligence, on the other hand, can be a process value for the future application of AI in global governance. Global collective intelligence is to address the question of the means to achieve global governance goals. Artificial intelligence can create a huge centrifuge effect. The development of AI may wash away the middle class, which was previously considered a stabilizer of society. The same possibility exists in the global community. The impact on global middle power means that a large number of middle-income countries could fall to the bottom or underdevelopment, and the growing accumulation of bottom power could lead to a revolt of the weak. In such a context, the formation of a tightly knit global community becomes even more relevant. Recent years have witnessed an increasingly fragmented, disunited, and even de-globalized or counter-globalized global society. The rapid development of artificial intelligence, on the other hand, may have a strong accelerating effect on this trend. Therefore, it is crucial to realize the dissemination and institutionalization of the concept of global collective intelligence.

References

1. Iji T (2017) The UN as an international mediator: from the post-cold war era to the twenty-first century. Glob Gov 23(1):83–100
2. Bazbauers A (2016) The world bank as a development teacher. Glob Gov 22(3):409–426
3. Mansbach R, Pirro E (2016) Putting the pieces together: international and EU institutions after the economic crisis. Glob Gov 22(1):99–115
4. Kacowicz A (2018) Regional governance and global governance: links and explanations. Glob Gov 24(1):61–79
5. Hobolt S (2016) The Brexit vote: a divided nation, a divided continent. J Eur Publ Policy 23(9):1259–1277
6. Honghua M (2017) China's approach to dealing with crisis and change in global governance. Soc Sci China (10). (in Chinese)
7. Van der Pijl K (2010) Competitors for western hegemony: past and present—as a global governance perspective. Contemp World (11). (in Chinese)
8. Qiqi G (2015) Global co-governance: the difference of world order between china and the west and its' reconciliation. World Econ Polit (4). (in Chinese)
9. Pichler H-K (1998) The godfathers of truth: max weber and carl Schmitt in Morgenthau's theory of power politics. Rev Int Stud 24(1):186
10. Zizek S (2001) On belief. Routledge, London, pp 143–144
11. Toynbee A (2005) Historical studies (trans by Beicheng L, Xiaoling G). Shanghai People's Publishing House, Shanghai, p 367. (in Chinese)
12. Stephen MD (2017) Emerging powers and emerging trends in global governance. Glob Gov 23(3):489–490
13. Lerner C (2017) The tower of babel revisited: global governance as a problematic solution to existential threats. N C J Law Technol 19(1):69–70
14. Russell S, Norvig P (2013) Artificial intelligence: a modern approach (3rd ed, trans by Jianping Y, En Z, Yue L et al). Tsinghua University Press, Beijing, p 5. in Chinese)
15. Matsuo Y (2018) Artificial intelligence frenzy: will robots surpass humans? (trans by Hanhong Z, Huabin G). Machinery Industry Press, Beijing, p 40. (in Chinese)
16. Kimber R (1991) Artificial intelligence and the study of democracy. Soc Sci Comput Rev 9(3):381–398
17. Barth T, Arnold E (1999) Artificial intelligence and administrative discretion: implications for public administration. Am Rev Public Adm 29(4):332–351
18. Duffy G, Tucker S (1995) Political science: artificial intelligence applications. Soc Sci Comput Rev 13(1):1–20

19. Jian H (2018) National strength based on big data: connotation and its assessment. Soc Sci China (6). (in Chinese)
20. Haixia Q (2018) Big data and innovation in international relations research. Soc Sci China (6). (in Chinese)
21. Qingling D (2018) Big data security situational awareness and conflict prediction. Soc Sci China (6). (in Chinese)
22. Qiqi G (2018) The risks of permanent marginalization: faced by developing countries and the mission of china in the age of artificial intelligence. Int Rev (4). (in Chinese)
23. Weiguang C, Jing Y (2018) Global governance of artificial intelligence: drivers, subjects, structures and mechanisms. Int Rev (4). (in Chinese)
24. Qingling D (2018) The new ethics of war: supervise and constrain the lethal autonomous weapons systems (LAWs). Int Rev (4). (in Chinese)
25. Chuanying L, Mallery J (2018) The process of global governance in artificial intelligence from the perspective of regime complex theory. Int Rev (4). (in Chinese)
26. Shuai F (2018) International relations in the age of artificial intelligence: towards a transformed and unequal world. Foreign Aff Rev (J Foreign Serv Inst) (1). (in Chinese)
27. Qiqi G (2017) China's special mission in the era of artificial intelligence. Explor Free Views (10). (in Chinese)
28. Hao L (2018) The combination of artificial intelligence and traditional industries deserves attention—'China artificial intelligence development report 2018' officially released. Sci Technol China (8). (in Chinese)
29. Matsuo Y (2018) Artificial intelligence frenzy: will robots surpass humans? (trans by Hanhong Z, Huabin G). Machinery Industry Press, Beijing, pp 186–187. (in Chinese)
30. Ezrachi A, Stucke M (2018) The algorithmic trap: super platforms, algorithmic monopolies, and scene deception (trans by Xiao Y). CITIC Press Group, Beijing, pp 315–316. (in Chinese)
31. Hansen HK, Porter T (2017) What do big data do in global governance? Glob Gov 23(1):36
32. de Siqueira IR (2017) Symbolic power in development politics: can 'Fragile States' fight with numbers?. Glob Gov 23(1):43
33. Best J (2017) The rise of measurement-driven governance: the case of international development. Glob Gov 23(2):163
34. Qiqi G (2018) A community of shared future for humanity and world politics in the era of artificial intelligence. Contemp World Social (3). (in Chinese)
35. Dos Santos T (1970) American economic association. Am Econ Rev 60(2):231
36. Amsden A (1992) Asia's next giant: south Korea and late industrialization. Oxford University Press, Oxford, pp 18–19
37. Brynjolfsson E, McAfee A (2016) The second machine revolution (trans by Yongjun J). CITIC Press Group, Beijing, p 252. (in Chinese)
38. Boyd D, Crawford K (2012) Critical questions for big data. Inf Commun Soc 15(5):673–674
39. Zechi Q, Shuqin Z, Shiding L (2016) From digital divide to dividend difference—a perspective of internet capital. Soc Sci China (10). (in Chinese)
40. Anderjevic M (2014) The big data divide. Int J Commun 8:1673–1689
41. Bostolom N (2015) Superintelligence: roadmap, dangers, and coping strategies (transl by Tiwei Z, Yuqing Z). CITIC Press Group, Beijing, pp 29, 64–67. (in Chinese)
42. Stuart Russell and Peter Norvig, *Artificial Intelligence: A Modern Approach (3rd ed.)*, p. 865.
43. Markoff J (2017) A brief history of artificial intelligence (trans by Xue G). Zhejiang People's Publishing House, Hangzhou, p 338. (in Chinese)
44. Clarke R (2011) Asimov's laws of robotics implications for information technology. In: Anderson M, Anderson SL (eds) Machine ethics. Cambridge University Press, NY, p 255
45. Allen C, Varner G, Zinser J (2000) Prolegomena to any future artificial moral agent. J Exp Theory Artif Intell 12(3):251
46. Sassoli M (2014) Autonomous weapons and international humanitarian law: advantages, open technical questions and legal issues to be clarified. Int Law Stud 90:339
47. Singer P (2016) Robot wars: the robotics revolution and rethinking in the 21st century (trans by Lu L, Yanan Z). Huazhong University of Science and Technology Press, Wuhan, p 305. (in Chinese)

48. Huntington S (1998) The clash of civilizations and the remaking of world order (trans by Qi Z et al). Xinhua Publishing House, Beijing, pp 3–5. (in Chinese)
49. Tee ZH, Ang L-M, Seng KP (2010) Smart guide system to assist visually impaired people in an indoor environment. IEEE Tech Rev 27(6):455
50. ten Kate J, Smitand G, Breedveld P (2017) 3D-prited upper limb prostheses: a review. Disabil Rehabil Assist Technol 12(3):300–314
51. Sen A (1992) Inequality reexamined. Harvard University Press, Cambridge, Mass, pp 39–40
52. Field M, Jette A (2007) The future of disability in America. National Academies Press, Washington DC, pp 22–25
53. Sénit C-A, Kalfagianni A, Biermann F (2016) Cyberdemocracy? Information and communication technologies in civil society consultations for sustainable development. Glob Gov 22(4):533
54. Kaplan J (2016) The age of artificial intelligence. In: The great future of wealth, work and thinking under human-machine symbiosis (trans by Pan L) Hangzhou: Zhejiang People's Publishing House, p 140. (in Chinese)
55. Qiqi G (2017) Global governance, mobility of people and community of shared destiny. World Econ Polit (1). (in Chinese)
56. Abrams D, Travaglino GA (2018) Immigration, political trust, and Brexit—testing an aversion amplification hypothesis. Br J Soc Psychol 57(2):310
57. Keping Y (2002) Introduction to global governance. Marx Rity (1). (in Chinese)
58. Floridi L (2018) Information ethics (trans by Ping X). Shanghai Translation Publishing House, Shanghai, p 429. (in Chinese)
59. Beitz C (1999) Political theory and international relations. Princeton University Press, Princeton, p 151
60. Pogge T (1989) Realizing Rawls. Cornell University Press, NY, pp 241–273
61. Pogge T (2004) The incoherence between Rawls's theories of justice. Fordham Law Rev 72(5):101–121
62. Wiener N (1978) The human use of human beings: cybernetics and society (trans by Bu C). The Commercial Press, Beijing, p 143. (in Chinese)
63. Müller V, Bostrom N (2016) Future progress in artificial intelligence: a survey of expert opinion. In: Fundamental issues of artificial intelligence, pp 553–571
64. Damnjanović I (2015) Polity without politics? Artificial intelligence versus democracy: lessons from Neal Asher's polity universe. Bull Sci Technol Soc 35(3–4):76–83
65. Wallach W, Allen C, Machines M (2009) Teaching robots right from wrong. Oxford University Press, New York, pp 3–11
66. Hussain A (2018) Ultimate intelligence: the future of perceptual machines and artificial intelligence. CITIC Publishing Group, Beijing, pp 48–49. (in Chinese)
67. Zhiyuan L (2016) Big data intelligence: machine learning and natural language processing techniques in the internet era. Electronic Industry Press, Beijing, pp 19–20. (in Chinese)
68. Qiqi G (2018) Artificial intelligence: taming the Severtan. Shanghai JiaoTong University Press, Shanghai, pp 150–151. (in Chinese)
69. Tuo C (2016) Global governance and national governance: two strategic considerations in contemporary China. Soc Sci China (6). (in Chinese)
70. Zhimin C (2016) National governance, global governance and the construction of world order. Soc Sci China (6). (in Chinese)
71. Xuelian L, Lu Y (2016) The global governance significance of national governance. Soc Sci China (6). (in Chinese)
72. Zhicheng W (2016) The impact of global governance on national governance. Soc Sci China (6). (in Chinese)
73. Zhenye L (2016) The interaction of global governance and national governance: ideological origins and realistic reflections. Soc Sci China (6). (in Chinese)
74. Qiqi G (2015) The theory and indexation of national participation in global governance. Soc Sci (1). (in Chinese)

75. Cate F, Kuner C (2012) Christopher Millard and Dan Jerker Svantesson, "The Challenge of 'Big Data' for Data Protection." Int Data Priv Law 2(2):49
76. Jinping X (28 Oct 2017) Secure a decisive victory in building a moderately prosperous society in all respects and strive for the great success of socialism with Chinese characteristics for a new era—report at the 19th national congress of the communist party of China. People's Dly. (in Chinese)
77. Wen G, Junrong T, Ke W (30 Jun 2014) Connecting Asia and Europe in trade and merging east and west civilizations: the grand concept of 'The Belt and Road Initiative' across time and space. People's Dly. (in Chinese)
78. Kaplan J, The age of artificial intelligence: the great future of wealth, work, and thinking under human-computer symbiosis, p 200
79. Qiqi G (2018) Life to death and eschatology: western ai pessimism and its critique. Study Explor (12). (in Chinese)
80. Fukuyama F (2017) Our post-human future: the consequences of the biotechnology revolution (trans by Lizhi H). Guangxi Normal University Press, Guilin, pp 216–217. (in Chinese)
81. Jinping X (1 Nov 2018) Xi Jinping stressed strengthening leadership in the ninth collective study of the Political Bureau of the CPC Central Committee to do a good job of planning clear tasks to tamp the foundation to promote the healthy development of China's new generation of artificial intelligence. People's Dly. (in Chinese)
82. Kang O (2018) The belt and road initiative' in the changing landscape of global governance. Soc Sci China (8). (in Chinese)
83. Jinping X (29 Mar 2015) Towards a community of destiny: creating a new future for Asia. People's Dly. (in Chinese)
84. Jinping X (2014) The governance of China. Foreign Languages Press, Beijing, p 262. (in Chinese)
85. Bin L (2017) The historical and practical dimensions of the new global governance consensus. Soc Sci China (10). (in Chinese)
86. Shipping T (2019) The changing international order and China's options. Soc Sci China (3). (in Chinese)
87. Weaver J (2018) Are robots people? (trans by Haian L, Teiying X, Qin X). Shanghai People's Publishing House, Shanghai, pp 181–183, 229. (in Chinese)
88. Acharya A (2016) The future of global governance: fragmentation may be inevitable and creative. Glob Gov 22(4):459
89. Acharya A (1997) Ideas, identity and institution-building: from the 'ASEAN Way' to the 'Asia Pacific Way.' Pac Rev 10(3):329
90. Floridi R (2016) The fourth revolution: how artificial intelligence is reshaping human reality (trans by Wenge W) Zhejiang People's Publishing House, Hangzhou, pp 203, 210. (in Chinese)
91. Alston P (2011) Lethal robotic technologies: the implications for human rights and international humanitarian law. J Law Inf Sci 21(2):59–60
92. Plesch D, Weiss TG (2015) 1945's lesson: 'Good Enough' global governance Ain't good enough. Glob Gov 21(2):203
93. Biermann F, Pattberg P, van Asselt H, Zelli F (2009) The fragmentation of global governance architectures: a framework for analysis. Glob Environ Polit 9(4):16
94. Patrick S (2014) The unruled world: the case for good enough global governance. Foreign Aff 93(1):58–73
95. Weiss TG, Wilkinson R (2018) The globally governed-everyday global governance. Glob Gov 24(2):193
96. Fox O, Stoett P (2016) Citizen participation in the UN sustainable development goals consultation process: toward global democratic governance? Glob Gov 22(4):555
97. Renn O (2017) Civic participation. In: Grenval A (ed) Handbook of technological ethics. Social Science Literature Press, Beijing, pp 696–702 (in Chinese)
98. Ayoub K, Payneb K (2015) Strategy in the age of artificial intelligence. J Strateg Stud 39(5–6):793–819

99. Huntington S (2005) Who are we? The challenge to American's national identity (trans by Kexiong C). Xinhua Publishing House, Beijing, pp 183–191. (in Chinese)
100. Rothblatt M (2016) The virtual man (trans by Xue G) Zhejiang People's Publishing House, Hangzhou, p 317. (in Chinese)
101. Yutaka Matsuo, "Artificial Intelligence Frenzy: Will Robots Overtake Humans?", p. 110.
102. Zhihua Z (2016) Machine learning. Tsinghua University Press, Beijing, p 113. (in Chinese)
103. Kaplan J, The age of artificial intelligence: the great future of wealth, work, and thinking under human-computer symbiosis, p 168

Chapter 4
Blockchain and the Transformation of Global Economic Governance

Due to the interference of multiple factors such as institutional inertia, path dependence, and governance hegemony, global economic governance is difficult to effectively adapt to the current development reality of global society. There is an apparent conflict between the current trend of multi-centralization and the previous centralized structure, which leads to the inherent flaws and expanding systemic risks in global economic governance. Among them, the original "center-periphery" international monetary pattern has not been broken, and the U.S. dollar is still dominant in the international monetary system. This international monetary pattern intensifies the competition between sovereign credit currencies and compresses the space for transforming and upgrading the global economic governance system. Blockchain technology can accelerate the transformation of global economic governance from centralized to multicentric by building a self-trusting ecology, a multifaceted integration mechanism, and an open collaboration system. To this end, we believe that we can build a set of E-SDR super-sovereign digital currency system based on blockchain technology with the basic structure of "digital currency system—digital financial account system -digital identity verification system" to create a new space for the reform of the international monetary system and the construction of a global justice economic order. There is still a considerable gap between developing and developed countries regarding blockchain technology's basic theory, industrial structure, and system construction. Therefore, China should seize the opportunity for technological development in time, accelerate research on blockchain theory and underlying technology, and promote open-source sharing of relevant developments. By doing so, China can realize new cooperation in blockchain and digital currency construction and contribute Chinese wisdom and solutions to global cooperation. This can help China lead multilateral cooperation on blockchain technology and digital currencies.

Q. Gao and J. Zhang, *Artificial Intelligence Governance and the Blockchain Revolution*, Artificial Intelligence and the Rule of Law,
https://doi.org/10.1007/978-981-99-9211-9_4

4.1 The Evolution of the Bretton Woods System and the Expansion of Dollar Hegemony

With the further development of globalization, the global economic governance system has undergone tremendous changes, and the transformation and upgrading of this system have become an important topic of discussion in the global society. From the perspective of development trends, the globalization of the world economy and the multi-polarization of politics has provided developing countries with a certain space for development and developing countries have provided a strong impetus for global development. It has gradually broken the original global power distribution pattern, and global economic governance has thus begun to show the realistic tendency of diversifying governance subjects and multi-centralizing governance regions. However, from the perspective of the actual structure, the developed countries still maintain the global economic pattern with themselves as the center and the developing countries as the periphery by their first-mover advantage and solid national power. They use this to strengthen the suppression and exploitation of the developing countries continuously. Therefore, the current global economic governance system cannot effectively adapt to the current development trend of global society, in which there is an apparent conflict between the current trend of multi-centralization and the previous centralized structure. This makes it impossible to effectively solve the new problems brought about by the development of global society, which leads to the expansion of the inherent flaws and systemic risks in the original global economic governance system.

In addition, from the perspective of technological development, the advancement of digital technology and the change of financial solutions provide a new perspective on global economic governance and an effective way to control the division and integration of power in the global economic order evolution. Among them, blockchain can realize the construction of polycentric architecture and distributed transactions through the use of cryptographic chain block structure, distributed nodes and consensus algorithms, and other technologies. Moreover, applying this technology will help accelerate the transformation of global economic governance from a single centralized to a polycentric one. To this end, this section aims to analyze the evolution of the international monetary and currency system after World War II, the current context of the transformation of global economic governance, and discuss in depth the current status of global economic governance as well as its possible future development direction. At the same time, this chapter also sorts out the nature, characteristics, and application levels of blockchain technology and its utility for global economic governance and proposes the idea of establishing an E-SDR super sovereign digital currency system with a view to providing a new perspective for promoting the transformation and upgrading of global economic governance.

An international currency is used for international trade, commodity valuation, and value storage when the legal tender of a sovereign country breaks through geographical and political boundaries.[1] So far, the international monetary system still maintains a certain degree of "center-periphery" structure, that is, the local currency of the "center countries" become the international currencies, and the "peripheral countries" exchange the right to use this international currency at the cost of international seigniorage.[2] At the same time, the "center countries," which has the dominant power over the international currency, designs and implements a series of institutional arrangements to influence the development of the international monetary system and the redistribution of global wealth in order to achieve the purpose of safeguarding its national interests. In addition, although the "peripheral countries" have taken a series of measures to change the system, they have not been able to effectively reform the international monetary system due to the hegemonic restrictions of the "central countries" and the cost of reform. To a certain extent, it has also greatly depended on the original "center-periphery" system.

In fact, since the creation of the Bretton Woods system, the U.S. dollar has gradually acquired a central position in the international monetary system; that is, the dollar has become the "central currency" closely surrounded by the "peripheral currencies" of other countries.[3] On the one hand, the dollar has become the intermediate anchor between gold and most national currencies; that is, the dollar is pegged to gold, and other countries' currencies are pegged to the dollar. This central position of the dollar has made it the most important payment currency for international trade and the most important reserve currency for the countries and regions of the capitalist system. On the other hand, based on the hegemonic position of the U.S. dollar, the U.S. has gradually gained international financial hegemony and discursive power, that is, the U.S. dominates the formulation and adjustment of global trade rules to ensure that

---

[1] Cooper [1].

[2] Gilpin [2].

[3] The establishment of the international monetary system after World War II was initially conceived in two ways, namely, the White Plan and the Keynesian Plan, with the main differences as follows: (1) The "White Plan" adopts the fund system, it would grant access to sovereign countries on the basis of their purchases, envisages the Unita as the unit of account (the actual circulation is still the dollar and other sovereign credit currencies), and takes gold as its currency anchor, and carries out the foreign exchange stabilization program of asymmetric adjustment, i.e., the deficit country bears the majority of the adjustment responsibility; (2) The "Keynesian Plan" adopts the banking system, with Bancor as the super-sovereign reserve currency (with global trade volume as its monetary anchor), and implements a symmetrically-adjusted foreign exchange stabilization program, with surplus and deficit countries proportionally sharing the adjustment responsibility. Theoretically, the "Keynesian plan" is more reasonable because it promotes a super-sovereign reserve currency, uses global trade volume as the anchor, and focuses on the relative fairness of rights and obligations in the balance of payments imbalance adjustment mechanism. However, as the dominant player in the construction of the post-war global economic pattern, the United States has vigorously pursued its priority position in the international monetary system. Therefore, the "Keynesian Plan" eventually fell through, and the "White Plan" became the blueprint for formulating the post-war international financial system. See Wu Xiaoling and Wu Ge, "'New Wyatt Plan' or 'New Keynesian Plan': How to Construct a Stable and Effective International Monetary System," *Exploration and Free Views*, No. 8, 2014, pp. 59–62 (in Chinese).

the global economic governance system always evolves in the direction favorable to the U.S. and to achieve economic plunder and control over other countries.

As the "world standard currency," the U.S. dollar has a high monopoly and circulation. Therefore, the U.S. can capture the international minting and inflation taxes on a global scale just by issuing dollars, i.e., to create external devaluation and internal inflation of the U.S. dollar by over-issuing dollars, thus reducing the pressure of U.S. external liabilities.[4] At the same time, the U.S. can adjust its monetary policy to promote its commodity exports and reduce its debt levels, while countries that use the dollar or acquire dollar assets will bear part of the burden of imbalance from this economic adjustment. In response, Benn Steil, director of the International Economics Department of the Council on Foreign Relations, believes that the Bretton Woods system is not only an international financial solution but also part of the U.S. geopolitical strategy, one of the core elements of which is to build a new international financial system to form a suppression of other countries.[5] In addition, the international monetary system itself is characterized by solid path dependence and continuous self-reinforcement, and this institutionalized support mechanism makes the dollar have a strong network effect in the international monetary system. As Barry Eichengreen, a professor of economics at the University of California, Berkeley, said, as long as the dollar has enough liquid financial markets, it can continue to maintain its position in the international monetary system, and it is difficult for other countries' currencies to break through the dollar-centric shackles in the international monetary system.[6]

Of course, although the Bretton Woods system did play an important role in promoting global post-war economic recovery and international trade growth, the dollar-centered international monetary system under its rule has natural instability and unsustainability. The most representative study on this issue is "Triffin's Dilemma," proposed by Robert Triffin, an American economist and professor at Yale University. He believes that under the Bretton Woods system, the U.S. dollar has a contradictory dual function, which is mainly manifested as follows: (1) in order to meet the needs of global economic development for the dollar reserves, the U.S. maintains the dollar liquidity by maintaining a continuous balance of payments deficit, but a long-term balance of payments deficit will lead to the devaluation of the dollar. (2) In order to keep the value of the dollar stable, the U.S. would have to maintain a long-term balance of payments surplus, but this, in turn, would result in insufficient liquidity for the dollar to provide sufficient monetary support for international settlements.[7] Therefore, this paradox between the two shows that the Bretton

[4] Wei [3].

[5] Steil [4].

[6] Eichengreen and Flandreau [5].

[7] Indeed, as long as the gold and foreign exchange reserve mechanisms operate in parallel, the gold stockpile will continue to be used as the ultimate repository of value for the monetary authorities, and instability in the international monetary system will not be eradicated. For a related study, see Robert Triffin, "Gold and the Dollar Crisis: The Future of Convertibility," *International Affairs*, Vol. 37, No. 1, 1961, pp. 251–258; Robert Triffin, "The International Role and Fate of the Dollar," *Foreign Affairs*, Vol. 57, No. 2, 1978, pp. 269–286.

Woods system has its own inherent contradictions that cannot be overcome. Meanwhile, Paul Krugman, a professor of economics at Princeton University, argued that under the Bretton Woods system, although the independence of national monetary policies and exchange rate stability were achieved, the international movement of capital was severely restricted. This means that the system did not break the monetary trilemma and, in fact, exacerbated the global capital liquidity crisis and the risk of international exchange rate volatility.[8] In addition, Francis Gavin, a professor of international affairs at the University of Texas, pointed out that the Bretton Woods system is inherently highly politicized, i.e., the United States needs to maintain the system more effectively through constant intervention and control. This has led to frequent monetary conflicts and even confrontations in the international monetary system caused by political interference.[9] Therefore, with the deepening of globalization and the intensification of the balance of payments crisis, the Bretton Woods system gradually began to collapse in the 1960s and 1970s.[10]

---

[8] The "monetary trilemma" mainly refers to the fact that a country can only achieve two of the three policy objectives of "monetary policy independence," "exchange rate stability" and "capital mobility," and it is not possible to achieve all three objectives at the same time. See Maurice Obstfeld, Marc Melitz, Paul Krugman, *International Economics: Theory and Policy*, New York: Pearson Addison Press, 2017, pp. 534–537.

[9] Gavin [6].

[10] The disintegration of the Bretton Woods system has experienced a number of dollar crises, which occurred in the following process: (1) After World War II, due to the U.S. in the "Marshall Plan" and "Dodge Plan" in the implementation of the "cheap money" policy, resulting in the deterioration of the U.S. balance of payments situation, the dollar depreciation pressure doubled. In October 1960, stimulated by the London gold market prices soared, the dollar then fell sharply and sold off on a massive scale, the first dollar crisis thus broke out. (2) In March 1965, the outbreak of the Vietnam war led to the United States international income and expenditure continued to deteriorate, the second dollar crisis broke out. (3) In August 1971, the U.S. House of Representatives International Trade and Balance of Payments Committee subcommittee issued a report that "the dollar is overvalued, suggesting the use of flexible exchange rate mechanism to solve the U.S. deficit problem." Nixon, then President of the United States, announced the implementation of the "New Economic Policy" and unilaterally closed the window for exchanging the dollar for gold, i.e., terminated the obligation to exchange the dollar for gold. This immediately led to a wave of dollar selling, the third dollar crisis also broke out. (4) In December 1971, the "Group of Ten" signed the "Smith Agreement," trying to rebuild the fixed exchange rate system by establishing a different ratio. However, due to the termination of the convertibility of the dollar and gold, and the member countries' currencies have realized the de facto free float, so the agreement did not eliminate the crisis of confidence in the dollar and failed to restore the original pegged exchange rate system. (5) In February 1973, as the U.S. balance of payments situation continued to deteriorate, the U.S. was forced to announce another devaluation of the dollar. (6) In March 1973, the countries of the Western European Common Market announced a joint floating system for the US dollar, while countries such as Britain and Italy implemented a separate floating system for the US dollar. The original gold-based, dollar-centered adjustable fixed exchange rate system was completely dismantled. Although the United States during this period also took such as the establishment of the gold treasury, the implementation of the "gold dual-price system" and "restraint withdrawal of gold agreement" and other means to maintain the ratio between the dollar and gold. However, the recurrence of balance-of-payments crises led to these measures did not solve the depreciation of the dollar pressure. (7) In January 1976, the international monetary system interim committee adopted the "Jamaica Agreement"; in April of the same year, the international monetary fund through the second amendment of the International

After the Bretton Woods system collapsed, the United States sought to construct a "hegemonic triangle" based on political, military, and technological dominance. Through this approach, the U.S. gradually achieved the dollarization of oil and other commodity denomination units. By influencing commodity trading prices and supply relations, the U.S. was able to strengthen the hegemonic position of the dollar. This means that the "gold dollar" is replaced by the "petro-dollar" valuation mechanism, and the global economic system has entered the "petro-dollar" valuation mechanism as the fundamental feature of the "Jamaican monetary system." The "Jamaican monetary system" was the post-Bretton Woods system period. In this period, the U.S. dollar unified the original gold pricing power and control of oil and other commodities production and trade, thus building a new basis of trust for itself and further enhancing the U.S. hegemonic position. Given this, David Spiro points out that the "petro-dollar" pricing mechanism allows the United States to influence or even manipulate the prices of commodities such as oil by changing domestic policies, thus achieving control over the distribution of strategic resources in the world and tying the interests of other countries to the United States.[11] However, while this new currency anchoring has mitigated the risk of a collapse of the international settlement system due to exchange rate fluctuations, it has also created new risks. McKinsey Global Institute researchers Diana Farrell and Susan Lund point out that the "petro-dollar" pricing mechanism has increased the size of "petro-dollar" sovereign wealth funds. The relatively high-risk appetite of these funds has contributed to the volatility of global capital markets.[12]

Of course, the basis for forming the "petro-dollar" system remains the United States itself hegemonic position. The U.S. economic strength and market size, the openness of financial markets, and a sound legal and political system strongly support the "petro-dollar" system. Nevertheless, more importantly, the "petro-dollar" pricing mechanism has contributed to forming the "petro-dollar-denominated financial assets" ring-fencing system. In order to alleviate the massive inflation caused by the collapse of the Bretton Woods system, the U.S. absorbed the overflowing dollars through direct and indirect ways such as adjusting interest rates, deregulating finance and increasing financial product innovation, and thus controlling the amount of dollar and maintaining the stability of the dollar.[13] At the same time, in the post-Bretton Woods period, the free flow of capital has become the basic international consensus. For the major reserve currency is single in essence, and for the need of liquidity and preservation of value, countries still first used U.S. bonds and other U.S. dollar assets as the main way to undertake their dollar surpluses.[14] As a result, the U.S.

Monetary Found Agreement, the agreement confirmed the non-monetization of gold, and formally recognized the freedom of member states in the exchange rate system. Thus, the Bretton Woods system, with the "gold-dollar" standard as its cornerstone, was formally dismantled. Mandel [7], Hogan [8], Eichengreen [9].

[11] Spiro [10].

[12] Farrell and Lund [11].

[13] Hudson [12].

[14] Guangbin et al. [13].

dollar in the global market flowed back to the U.S. in the form of U.S. dollar assets in large quantities, and the U.S. dollar also became a store of value for countries in disguise in the form of various financial assets. In this process, the widespread and convenient use of the U.S. dollar and the stability of the U.S. comprehensive power have prompted the U.S. national credit to gradually become a new type of anchor for the U.S. dollar. The formation of this system eliminated the need for the dollar to rely on the scarcity of anchors to secure its credit. The dollar thus became an international currency based entirely on unilateral credit.[15]

Therefore, although the post-Bretton Woods system formed a pluralistic monetary system, it was still essentially a dollar-based credit monetary system, and the currencies of most countries (tiny and medium-sized developing countries) were still mainly anchored to the dollar in the international market.[16] At the same time, as the dollar's substance and role increased, the United States adopted a more assertive monetary policy and used it to shape and expand the U.S.-centered global system. As Robert Keohane, a professor of political science at Duke University, and Joseph Nye, a professor of political science at Harvard University, have argued, when the demands of currency exchange no longer constrain the United States, it can fully use its economic, political and military power to influence the rules of the international monetary game.[17]

In fact, when faced with conflicting national interests and international obligations, the U.S. prefers to formulate policies consistent with its own national interests. Under the post-Bretton Woods system, the U.S. dollar remains the main reserve asset and de facto pegged currency for global trade surplus countries. However, the U.S. is not responsible for maintaining exchange rate stability. As a result, the U.S. can choose its fiscal and monetary policies more autonomously and avoid the enormous economic risks associated with continued exchange rate volatility. Michael Hudson, a professor of economics at the University of Missouri, believes that the U.S. has formed a "Super Imperialism" in the form of state capitalism based on the hegemonic position of the U.S. dollar. The "Plaza Accord" and the "Louvre Agreement" in the 1980s and 1990s are the most typical illustrations.[18] However, other countries are still facing the risks that may result from the centralization of the dollar. In this regard, Ronald McKinnon, a professor of economics at Stanford University, argues that continuing the dollar hegemony has made it impossible for most countries to construct their foreign claims in their currencies. In contrast, those countries whose current account surpluses have led to increasing foreign claims continue to face the "The Conflicted Virtue" dilemma.[19]

---

[15] Cooper [14].

[16] Xiangsui [15].

[17] Keohane and Nye [16].

[18] According to Hudson, the main manifestation of U.S. "Super Imperialism" is the U.S. suppression and exploitation of other countries through the centralization of the U.S. dollar, intervention in central banks, and the manipulation of multilateral institutions of intergovernmental capital. See Hudson [17].

[19] McKinnon believes that the hegemonic position of the U.S. dollar prevents East Asian countries from lending in their own currencies, only the U.S. dollar can be formed as the basic settlement

The U.S. is actively promoting the development of international financial derivatives markets to stimulate global demand for dollar transactions, investments, and reserves, thereby consolidating and expanding the hegemonic position of the dollar. This means that the U.S. can dominate global capital flows and contraction of trade activities through monetary and fiscal policy adjustments and thus seize the huge wealth of manufacturing and resource countries.[20] This has led to the "Dollar Trap" dilemma in the international monetary system. In response, Eswar Prasad, a professor of economics at Cornell University, points out that because a large amount of global financial assets exist in the form of U.S. dollars, the United States has become the leading supplier of safe global financial assets in disguise. Against the backdrop of significant growth in international reserves and frequent financial crises, the international community's demand for safe assets has become an additional force supporting the hegemonic position of the dollar.[21] In addition, the U.S. has gained great discourse power in the international financial system by virtue of the hegemonic position of the U.S. dollar and has seized substantial economic and political benefits in formulating or modifying the rules for handling international economic affairs.[22]

However, this cyclical change in the dollar has exacerbated the volatility of the international monetary system and has had a significant impact on global capital flows and asset allocation, with some international economic activities even suffering a huge shock (Fig. 4.1).[23] Specifically, the spillover effects of the cyclical changes in the dollar are divided into four aspects. First, from the perspective of internal and external financing, the phased weakness of the dollar has forced countries with unsynchronized economic cycles to respond to the liquidity crunch by raising interest rates. This has led to a significant increase in the cost of internal financing and external debt levels in these countries. Countries that are overly dependent on foreign capital and have excessive short-term debt are highly vulnerable to a debt crisis under this

currency for external debt. It partly increases the pressure of passive appreciation of their domestic currencies to a certain extent. At the same time, East Asian countries tend to be able to maintain a trade surplus in foreign trade, their current account can maintain a high level of surplus. However, the external claims of these countries are also increasing so that the debtor countries gradually began to complain about the bilateral trade deficit caused by the undervaluation of the East Asian countries' local currencies, thus further increasing the expectations of holders of domestic dollar assets for the appreciation of the local currency. However, if East Asian countries' currencies appreciate, their domestic dollar asset holders will suffer losses; if not, East Asian countries will face more complaints and greater pressure from debtor countries. Therefore, East Asian countries are thus caught in the dilemma of "there can be neither appreciation nor non-appreciation." See Mckinnon and Schnabl [18].

20 Xiangsui [19].

21 "Currency Trap" refers to a series of financial dilemmas faced by one or more countries as a result of their heavy reliance on the U.S. dollar for their economic activities. Prasad believes that in the post-financial crisis era, due to the lack of suitable substitutes for safe assets, the world's official investors and private investors passively increased their dependence on financial assets denominated in U.S. dollars. It is this increased dependence that has led to the strengthening of the hegemony of the dollar. See Prasad [20].

22 Yulu [21].

23 Maurice [22].

scenario. In this regard, Mansoor Dailami, head of international finance at the World Bank's Global Development Forecast Bureau, and Paul Masson, professor at the Rotman School of Management at the University of Toronto, Canada, point out that expectations of monetary policy changes in the U.S. have a significant impact on emerging market spreads and, to some extent, exacerbate the global spread of debt crises—the spread of the debt crisis on a global scale.[24] Second, from the perspective of international trade, a period of a strong dollar benefits other countries' exports. However, it will lead to a relative depreciation of other countries' currencies and an increase in the cost of imports. This may lead to a sustained increase in prices in other countries, which in turn may expose them to the risk of imported inflation. And those countries in the relatively downstream position of the value chain link will face the risk of deterioration of foreign trade in the short term. According to Krugman, although the strengthening of the dollar is beneficial to other countries' exports, these countries still face the risk of deterioration of foreign trade in the short term due to their relatively downstream position in the value chain and the influence of factors such as resource export orientation.[25] Third, from the perspective of currency liquidity, under the post-Bretton Woods system, as most countries implemented a floating exchange rate mechanism, so once the dollar strengthened, it would cause a continuous loss of foreign exchange reserves of these countries in dollars. As a result, the central banks of these countries were forced to tighten their base money allocations, and their domestic monetary liquidity contracted as a result. This has led to a structural shortage of global liquidity and the accompanying deflation, exacerbating the contradiction between supply and demand for international reserve currencies. As Michael Dooley, a professor of economics at the University of California, Santa Cruz, put it, the floating exchange rate system and the expansion of capital markets actually contributed to the "Bretton Woods recovery," i.e., the formation of a new "center-periphery" structure of the international monetary system.[26] Fourth, regarding asset allocation, the phased weakness of the U.S. dollar has caused a large amount of excess liquid capital to flood into emerging markets. Broad asset prices in emerging market countries, such as stocks and real estate, rose rapidly. However, once the dollar resumed its strength, liquid capital flowed back to the U.S. This process accelerates the bubbling of domestic assets and excessive dollar liabilities for "fragile countries" that are more sensitive to liquidity and interest rates. The original asset bubble and excessive dollar indebtedness could easily lead to a domestic economic crisis.[27]

---

[24] Dailami et al. [23].

[25] Krugman [24].

[26] Dooley et al. [25].

[27] The term "fragile states" is derived from the Fragile Five concept developed by Morgan Stanley in 2013, which refers to emerging countries that are economically overly dependent on foreign investment, whose currencies are highly susceptible to fluctuations in the U.S. dollar, and who are vulnerable to risk. The indicators include the current account balance, the ratio of foreign reserves to external debt, foreign investor holdings of government bonds, dollar-denominated debt, inflation, and real interest rate differentials. See Umit [26].

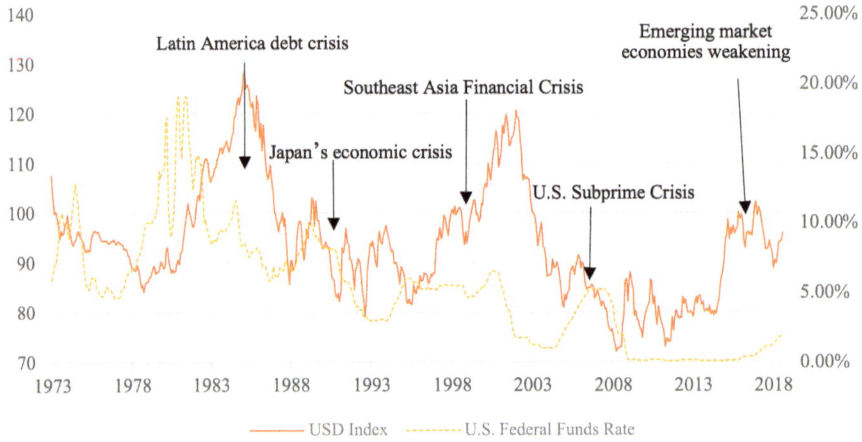

Fig. 4.1 U.S. monetary policy, the U.S. dollar index and the post-World War II global financial crisis. *Source* Investing, DXY, FFR, https://cn.investing.com, accessed on December 12, 2018

Based on the hegemonic position of the U.S. dollar in the international monetary system, the U.S. constructed a global cross-border payment and settlement system dominated by the U.S., which provided the necessary technical guarantee and basic conditions for the U.S. to impose financial sanctions. In 1977, the Society for Worldwide Interbank Financial Telecommunication (SWIFT) was formally established under the leadership of large banks from the United States and some European countries, and has since become a globally dominant international payment clearing system. As an international unified inter-bank payment information transmission system, SWIFT has built a unified information channel for global monetary fund flow, and its message standard format and transaction mode have become a standardized process for banks to exchange data. Susan Scott, a professor at the London School of Economics and Political Science, believes that SWIFT has formed a core part of the international financial services infrastructure and even once became a monopoly in the international interbank cross-border exchange market.[28] However, despite its nominal role as an international inter-bank non-profit international cooperation organization, SWIFT is a cross-border funds clearing system established by the United States based on the hegemony of the U.S. dollar.[29] On the one hand, the U.S. not only dominates the establishment of SWIFT but also de facto controls the operation of SWIFT by uniting with its traditional allies to occupy a majority of the board seats. On the other hand, through the monopoly of SWIFT data and its management, the U.S. cannot only grasp information on global financial flows effectively but also impose financial sanctions on other countries by restricting access

[28] Susan Scott and Zachariadis [27].

[29] SWIFT has a Board of Directors with 25 seats, which rotate every three years. The United States, the United Kingdom, France, Germany, Belgium, and Switzerland each have two seats on the Board, the other core Member States have one seat, and the majority of ordinary Member States have no seat. Susan and Zachariadis [28].

to international payment settlements.[30] Sylvie Peyrou, a researcher at the Center for European Studies at the University of Pau, France, believes that the U.S. use of SWIFT data goes far beyond the normal scope and threatens and de facto initiates sanctions against other countries by manipulating SWIFT.[31] SWIFT is only part of the "financial infrastructure" built by the U.S. based on the hegemony of the U.S. dollar in global payment settlement and financial products trading, which is a derivation of the power of the U.S. dollar to gain dominance in the international monetary system. The formation of this power structure not only strengthens the international status of the U.S. dollar but also leads to other countries being forced to strengthen their dependence on the U.S. dollar.

Of course, despite facing the challenges of currencies such as the euro and the yen and the constraints of the international system, the dollar still maintains an overwhelmingly dominant position in global trade and financial activities. On the contrary, the rise of emerging currencies has instead accelerated the fragmentation of international monetary power and the reorganization of power structures.[32] Jonathan Kirshner, a professor of international political economy at Cornell University, points out that while the breakup of the Bretton Woods system has weakened the dollar's power in the outer zone, the U.S. has become more carefree in its use of currency manipulation tools. The new international monetary system has given some countries access to the implementation of monetary power. This has intensified the competition and conflict in the international monetary system.[33] Meanwhile, Lawrence Summers, a professor of economics at Harvard University, argues that under the dollar-dominated international monetary system, the vast capital export and twin deficits (fiscal deficit, trade deficit) taken by the U.S. make it the largest market in the world, the dependence of emerging countries' economic development on U.S. demand is also rising. In this pattern, emerging countries maintain the existing "equilibrium" out of cost avoidance while the U.S. finances its trade deficit with external capital inflows. This leads to a "Balance of Financial Terror" in the international financial landscape.[34] In addition, Guillermo Calvo, a professor at the University of

---

[30] For example, in September 2005, the United States found that there were transactions between Banco Delta Macau and the Democratic People's Republic of Korea after analyzing data from SWIFT and the New York Clearing House Interbank Payments System. Based on this finding, the U.S. Treasury Department sanctioned Banco Delta Macau, which had multiple North Korean clients. The ban has also become an important weight for the U.S. in the six-party talks on North Korea's nuclear program. Meanwhile, the U.S. National Security Agency (NSA) has always monitored SWIFT data information, according to documents released by Edward Snowden in 2013. In addition, in May 2018, the U.S. announced its withdrawal from the Iran nuclear deal and restarted economic sanctions against Iran, which also required SWIFT to exclude Iranian banks. See Lee Mathews, "Shadow Brokers Leak Reveals The NSA's Deep Access Into SWIFT Banking Network," FORBES NEWS, April 15, 2017, https://www.forbes.com/sites/leemathews/2017/04/15/shadow-brokers-leak-reveals-the-nsas-deep-access-into-swift-banking-network/#77ddd0a4445b.

[31] Peyrou [29].

[32] Yuyan and Jingchun [30].

[33] Kirshner [31].

[34] Summers believes that the continued expansion of the U.S. twin deficits has led to greater pressure on the United States to service its debt. If emerging market countries do not continue to buy dollar

Maryland, and Carmen Reinhart, a professor of finance at Harvard University, argue that the emergence of multiple currencies and floating exchange rate mechanisms have forced sovereign states to reduce exchange rate volatility through government intervention, leading to the "fear of floating" was created.[35]

In general, the weakening of the macroeconomic power of the United States is an important reason for the collapse of the Bretton Woods system. However, the separation of the dollar from gold's fixed-value commitments and obligations has, on the contrary, allowed the U.S. dollar to enter a relatively unaccountable realm of freedom. The U.S. has thus completely freed itself from the restrictions imposed by the original mechanism on exercising U.S. economic power. It can thus choose and implement a more independent fiscal and monetary policy. Therefore, a strong dollar has become the inevitable choice for the U.S. monetary strategy and geopolitics. This is also an important reason why the "national interest orientation" in the U.S. global hegemony is relatively prominent. In addition, the end of the Cold War and the deepening of globalization have led to the gradual integration of the international political and economic systems, thus creating a favorable environment for the renewed global expansion of the U.S. dollar. However, while enjoying the benefits of dollar hegemony, the United States failed to fully assume its responsibility to maintain and coordinate the international monetary order. This has led to the increasing contradiction between the expanding financial asset bubble and the limited dollar money supply, the increasing international financial systemic risks, and the rising fragility of the international monetary system.

4.2 Transformation Scenario: Traditional Centrality and Realistic Polycentricity

The developed countries, by virtue of their first-mover advantage and national power, have long maintained such a pattern in global economic governance—a global economic pattern with themselves at the center and the developing countries at the periphery—and have used it to intensify their suppression and exploitation of

assets at this time, it is very likely to lead to the collapse of the dollar and trigger an economic crisis in the United States, which in turn negatively affects their own economic development. For this reason, emerging market countries have had to buy dollar assets. See Weixing and Guannan [32].

35 Under the floating exchange rate mechanism, some emerging market countries often experience large fluctuations in their international reserve rates due to a lack of international credibility. This not only has a great impact on their imports and exports but is also very likely to lead to domestic inflation. As a result, these countries have a fearful attitude towards exchange rate fluctuations, and tend to limit them to a small range through government intervention, leading to a large difference between the nominal and real exchange rates. See Calvo and Reinhart [33].

the developing countries.[36] However, against the backdrop of deepening globalization and rising levels of productivity in society, developing countries and emerging markets have grown rapidly. In contrast, the strength of traditional developed countries has been weakened, and the gap between the two has been narrowing. This makes global economic governance begin to show the development trend of diversified governance subjects and polycentric governance.

The contribution of developing countries and emerging markets to global economic growth is increasing and gradually changing the original pattern of world economic power. In 1980, developing countries and emerging economies accounted for 36.66% of world GDP in purchasing power parity (PPP), while developed markets accounted for 63.34%; in 2002, the GDP share of developing countries and emerging economies exceeded that of the Group of Seven (G7) for the first time; in 2008, the GDP share of developing countries and emerging economies rose to 51.21%, surpassing the 48.79% of developed countries for the first time; as of December 2018, the GDP share of developing countries and emerging economies has reached 59.77%; according to IMF projections, in 2020, the GDP share of developing countries and emerging economies will reach 60. 50%, and the share of developed countries, on the other hand, will decline to 39.50% (Fig. 4.2).

Second, developing countries and emerging economies have contributed to most global economic development. From 1980 to 2017, the average annual growth rate of the global economy was 3.49%, and the average growth rate of developing countries and emerging economies was 4.53%, among which the growth rate of Asian emerging economies was as high as 7.36%; the growth rate of developed countries was 2.42%, among which the core G7 countries grew at 2.22% (Fig. 4.3). The emerging economies led by China have made a huge contribution to the growth of the global economy. At the same time, according to the World Bank report, 150 of its 189 member countries are developing countries, and 108 of them have become "middle-income countries," i.e., have entered the so-called middle-income stage. At present, the combined GDP of these developing countries accounts for only one-third of the world's GDP, while their total population accounts for 74.32% of the global

[36] The "center-periphery" is the basic structure used by the dependency theory to analyze the global politics and economy. According to the Argentine economist Raúl Prebisch, developed and developing countries, or center and periphery, are in an unequal developmental position in the world economy, and the asymmetry of the international division of labor deepens this "center-periphery" relationship. See Prebisch [34], Prebisch [35]; Egyptian economist Samir Amin argues that the expansion of capitalism has led to the formation of a "center-periphery" structural world system, where the capitalism of the "central state" has created a global oligarchy through monopolization and control of technology, natural resources, finance, global media, and means of mass destruction, see Amin [36], Amin [37]; Peter Preston, a professor of political science and international relations at the University of Birmingham, argues that the developed countries at the center of the dominant global structure are interconnected with the developing countries at the periphery through unequal exchanges. In contrast, the peripheral countries are heavily dependent on the technology, capital, and markets of the central countries, and this "center-periphery" model results in the peripheral countries remaining underdeveloped, see Preston [38].

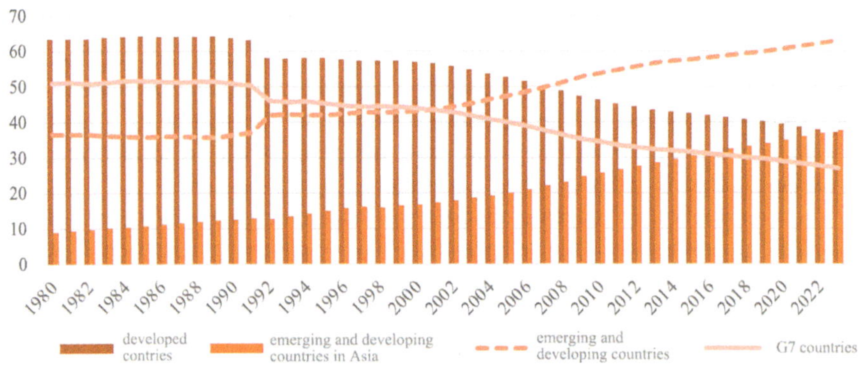

Fig. 4.2 Comparison of Global GDP Aggregate Shares. *Source* International Monetary Fund database, GDP Based on PPP (Share of the World) dataset, http://www.imf.org/external/datamapper, accessed December 13, 2018 (*Note* 2018–2022 is forecast data)

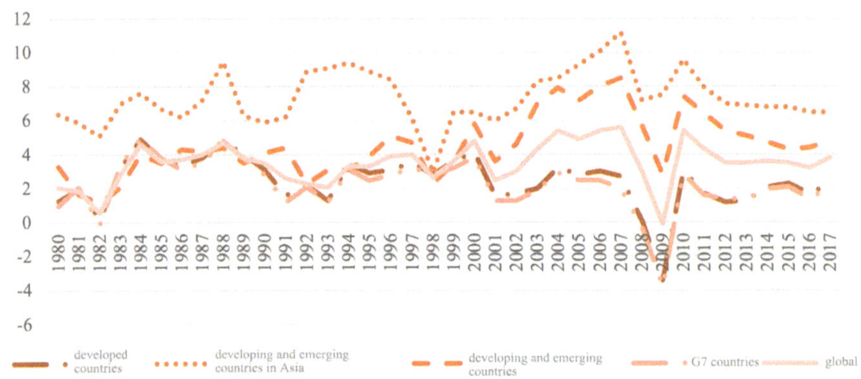

Fig. 4.3 Comparison of changes in global GDP growth rates. *Source* International Monetary Fund database, Real GDP Growth dataset, http://www.imf.org/external/datamapper, accessed on December 13, 2018

population.[37] Therefore, although most developing countries are still in the early stages of modernization, whether they can continue to move toward intermediate and advanced modernization in the future based on the current foundation, these developing countries will undoubtedly provide a strong follow-up growth engine for the development of the global economy.

Thirdly, developing countries and emerging markets are constantly improving their position in global trade and accelerating their integration into the global market. The division of labor in global value chains and the adjustment of global industrial layout make multinational companies continuously move their production bases

[37] United Nations Conference on Trade and Development, *World Investment Report 2018* http://worldinvestmentreport.unctad.org/world-investment-report-2018/#key-messages, September 13, 2018.

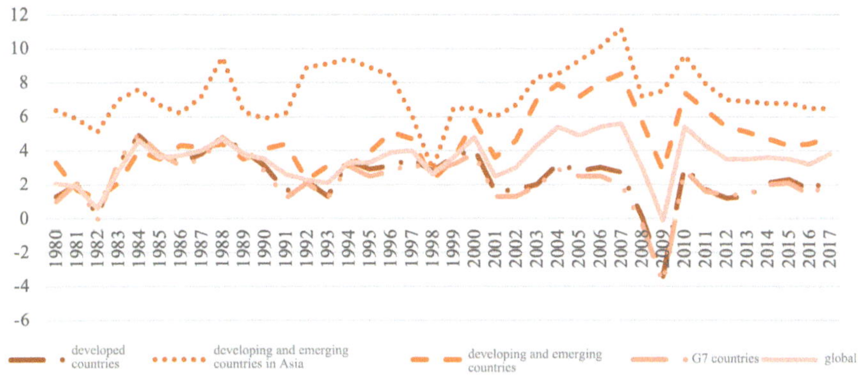

Fig. 4.4 Comparison of Global Current Account Balances. *Source* International Monetary Fund database, Current account balance U.S. dollars dataset http://www.imf.org/external/datamapper, accessed December 13, 2018

to emerging countries. In this transfer process, developing countries and emerging markets have become "world manufacturing factories" and "production workshops" for MNCs with abundant labor and resource advantages. The exports of developing countries and emerging markets have been multiplying, their share in international trade has been expanding, and their current account balances have been on the rise. Some developing countries have become important global trade surplus and creditor countries (Fig. 4.4). The formation of global value chains has increased the role of emerging economies in the global economy and enabled these countries to transform their own industries in industrial undertaking.[38]

Fourth, the international monetary system has begun to trend toward a pattern dominated by multiplc international currencies, with the emergence of "one super power and multi-great power" changes. By the third quarter of 2018, the share of the U.S. dollar in official foreign exchange reserves fell to 61.94% (the seventh consecutive quarterly decline) from 62.40% in the second quarter of 2018, the lowest share since the fourth quarter of 2013 (Fig. 4.5). Meanwhile, according to a SWIFT report released in December 2018, the share of the U.S. dollar in international payment usage in November was 39.56%, while the euro, the British pound, and the Japanese yen were about 34.13%, 7.27%, and 3.55%, respectively, while the Chinese yuan accounted for about 2.09%.[39] A diversified international monetary system is gradually taking shape, and this forms a stronger constraint on each issuing country of the international reserve currency and poses a strong challenge to the hegemony of the U.S. dollar. In this regard, Thomas Costigan, a researcher in the Department of International Political Economy at the University of Western Sydney, Australia, believes that the reserve and payment currencies in the current international monetary system

---

[38] Marchi [39].

[39] SWIFT, "International Payment," http://www.swift.com/products_services, accessed December 21, 2018.

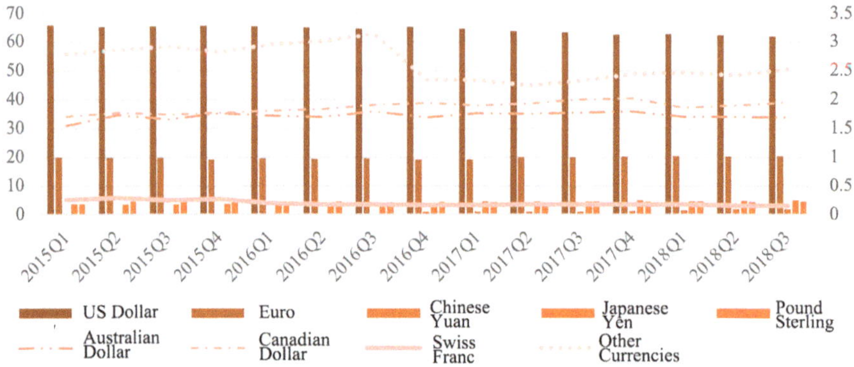

Fig. 4.5 Changes in the share of global foreign exchange allocated reserves. *Source* World Currency Composition of Official Foreign Exchange Reserves, http://www.imf.org/external/datamapper, accessed on December 13, 2018

are gradually decentralized, thus forming a more diverse and competing international monetary structure.[40]

However, due to the interference of multiple factors such as institutional inertia, path dependence, and governance hegemony, the incompatibility between the real development of global society and global economic governance has become more and more obvious, i.e., there is an obvious conflict between the current trend of multi-centralization and the previous centralized structure, and the inherent flaws and systemic risks in global economic governance have been expanding. This makes it impossible to effectively solve the new problems brought about by global social development, and the global economy faces problems such as insufficient momentum for subsequent development, frequent international financial chaos, and the rise of the wave of counter-globalization. The recent rise of unilateralism and trade protectionism illustrates this problem. Specifically, there are several problems in global economic governance as follows:

First, there needs to be more rationality in the distribution of power in global economic governance relative to the contribution of global development. The contribution of developing countries and emerging market countries to global economic growth has been increasing. However, these countries still need to be given the matching power of governance, especially since most small and medium-sized developing countries are excluded from the core of global economic governance. For example, the development of Southeast Asian countries after the 1997 Asian financial crisis can be seen as a development on the periphery of the capitalist world system.[41] At the same time, although developing countries have made their voice increasingly heard in global economic governance through their advanced national ability, they still need to gain relatively equal status with developed countries in formulating key rules and core standards. The developed countries, represented by the United States,

[40] Costigan [40].

[41] Zhengyi [41].

still hold the power to set global economic governance mechanisms and international rules.[42] In addition, the relative advantages gained by emerging market countries and developing countries come at the cost of massive inputs and consumption of labor, natural resources, and the environment. Those developing countries whose reforms have yet to show results and whose domestic economies are more vulnerable are instead facing more and more nascent risks.[43] Thus, the imbalance in the power structure of global economic governance has also led to a continuous increase in "Global Risk Sentiment" and "Risk Premium."[44]

Second, the original centralized global economic governance mechanism cannot effectively adapt to the current polycentric reality of global interconnection. Under the impetus of economic globalization and trade liberalization, globalization continues to deepen and integrate most countries and regions, which makes global economic governance issues expand and the governance field's boundaries increasingly blurred.[45] However, the original global economic governance system has been more global in perspective, i.e., it has been centralized in its approach to global economic development, has failed to reconcile the multiple forces emerging from globalization effectively, and has not been sufficiently linked to regional economic governance and domestic economic governance in each country. Susan Strange has argued that the plurality of subjects in international political and economic relations is an important component of the basic power structure and that the absence of interaction between these units in the power structure exacerbates the disorderly nature of the system.[46] Therefore, under the framework of such a centralized governance system, it is difficult to achieve multilateral consultation on governance policies among countries, and it is also difficult to form an effective consensus in the short term.[47] At the same time, the issues of global economic governance evolve dynamically with global development. Diversifying governance subjects and deepening stakeholder hierarchy also lead to the increasing objects and scope of global economic governance. In contrast, the centralized governance system lacks sufficient freedom, flexibility, and inclusiveness, especially considering the interaction between developed and emerging powers, and thus cannot keep pace with the changes of global economic governance. This makes it challenging to keep up with the pace of change in global economic governance.[48] In addition, technological change and financial innovation have led to the increasing complexity of the global production system and economic governance, and more and more non-governmental actors with

---

[42] Zhicheng [42].

[43] Keping [43].

[44] Global Risk Sentiment refers to the degree to which market participants hold a pessimistic outlook on the global economy. This sentiment is often transmitted ahead of actual economic developments. Risk Premium is the portion of return above the risk-free return. The higher the risk premium, the greater the relative risk. See Ferson and Harvey [44].

[45] Polanyi [45].

[46] Strange [46].

[47] Tuo [47].

[48] Xiujun [48].

technological power and private authority are involved in the operation of the global value chain, and it is difficult for the original global economic governance system to regulate these actors effectively.

Third, the fragmentation of global economic governance has been exacerbated by the development of global economic differentiation, which has reduced the effectiveness of global economic governance. The fragmentation of global economic governance is firstly reflected in the complexity and redundancy of governance mechanisms. This has created several mechanism complexes, leading multinational enterprises to pay many compliance costs under the constraints of multiple international and domestic rules.[49] With relatively small scale, those companies in the growth stage pay even higher costs for this. Michael Burda, professor of economics at Humboldt University in Berlin, Germany, argues that while the social division of labor and cost competition has increased the efficiency of social production, this has also led to the fragmentation of international trade rules and high systemic costs.[50] At the same time, according to the World Investment Report 2018 on global investment policy trends, more countries will strengthen their scrutiny of foreign investment in the future, and the share of global regulatory and restrictive investment policies will thus continue to increase.[51] In fact, due to the different concepts of economic governance and the state of the economy, there are large differences in countries' perceptions on core governance elements such as benefit distribution, cost burden, and rule mechanisms. As a result, it is often difficult for countries to reach an effective consensus on global economic governance, which in turn exacerbates the inefficiency of global economic governance.[52] In addition, against the background of the current restructuring of global economic governance rules, the implementation of policies agreed under the global economic governance system and their effectiveness varies. Because some rules fail to fully consider the multilateral situation and the differences between different regions, they impede the development of the global economy and even intensify the rise of trade protectionism.

Fourth, the problem of hegemonic dominance in global economic governance still exists, which makes global economic governance fall into the dilemma of involution and path dependence.[53] Although the original hegemonic governance model has

[49] Fan et al. [49].

[50] Burda and Dluhosch [50].

[51] United Nations Conference on Trade and Development, *World Investment Report 2018*, http://worldinvestmentreport.unctad.org/world-investment-report-2018/#key-messages, accessed October 13, 2018.

[52] Yu [51].

[53] Originally proposed by historical anthropologist Clifford Geertz, "involution" refers to the stagnation of a social or cultural pattern after it has reached a certain definite form. Involvement in global economic governance, on the other hand, refers to the fact that global economic governance tends to become self-reinforcing in a set direction, and is caught in a dilemma of transformation and upgrading, due to the presence of economies of scale, learning effects, coordination effects, adaptive expectations, and vested interest constraints. See Clifford Geertz, "Agricultural Involution: the Process of Ecological Change in Indonesia," *American Anthropologist*, Vol. 70, No. 3, 2010, pp. 599–600.

become unsustainable due to the relative decline and lack of legitimacy of the hegemonic powers themselves and the rise of a group of emerging powers, the embedding of developing countries in the global value chain has brought them closer to the developed countries.[54] In this process, as developing countries are in the middle and lower reaches of the value chain, once the macroeconomic policies of developed countries are adjusted, developing countries will face greater crises such as capital outflow, the pressure of currency devaluation, and global trade protectionism.[55] Meanwhile, emerging developing countries still need a more stable international development environment, and it is still difficult for emerging developing countries to bear the institutional and behavioral costs of promoting the transformation and upgrading of global economic governance. As a result, developing countries (especially small countries) want to promote the reform of governance mechanisms on the one hand. However, on the other hand, they can hardly afford the cost of institutional change and cannot promote the reform of the governance system.[56] As a result, developing countries are forced to choose to maintain the original governance mechanism and are trapped in path dependence. Additionally, developed countries have not promptly assumed governance responsibilities. Instead, they pursued a "unilateral logic" mindset, adhered to "Western standards" as an evaluation mechanism, and abused their governance power to control and plunder developing nations.[57]

In general, with the further deepening of globalization and the rise of developing countries and emerging market countries, global economic governance has entered an era of polycentric governance regions and pluralistic governance subjects. However, under the role of the original "center-periphery" global economic order, the majority of developing countries are still economically constrained by and dependent on developed countries. In contrast, developed countries in the center still have absolute advantages in important areas. This imbalance limits the space for holistic global development and restricts the effective implementation and upgrading of global economic governance, thus plunging the global economy into the double dilemma of development deficit and governance deficit. However, economic structures must be compatible with political realities, and a stable economic structure must be based on a shared political understanding.[58] Therefore, it is necessary to construct a global economic governance system with basic symmetry of power and responsibility and the development of common interests through reforms (Fig. 4.6).

---

[54] Zhimin [52].

[55] Zhengyi [53].

[56] Yaqing [54].

[57] Jing [55].

[58] Kissinger [56].

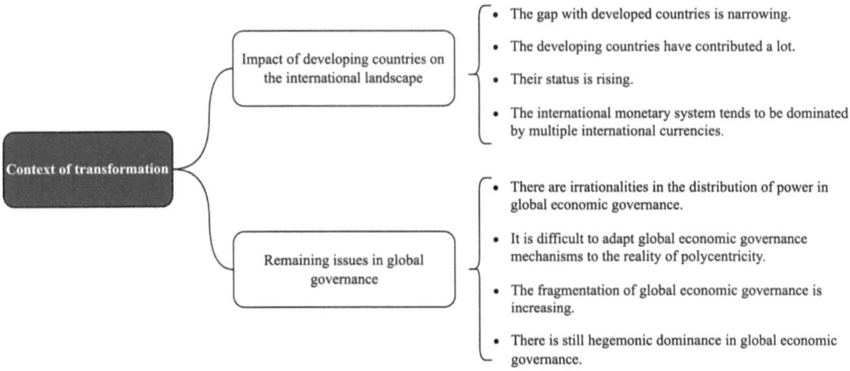

Fig. 4.6 Transformative contexts and salient issues in global economic governance

4.3 Technological Features of Blockchain and Its Significance for Global Economic Governance Transformation

Technological advances and changes in financial solutions in the digital era have provided a new perspective for the transformation of global economic governance. As a framework underlying technology, blockchain can utilize cryptographic chain block structure, peer-to-peer network, distributed algorithm, and data storage in the cloud to realize the construction of polycentric architecture and distributed network, and its core purpose is to create a set of "rules of the game" that is open, transparent, safe, reliable, efficient and intelligent. The application of this technology will effectively promote the transformation and upgrading of global economic governance.

Blockchain technology was first introduced as the underlying technology of Bitcoin in 2008 by Satoshi Nakamoto in his seminal paper, "Bitcoin: A Peer-to-Peer Electronic Cash System."[59] In January 2009, the launch of the Bitcoin network officially marked the beginning of blockchain applications. Currently, there is a consensus definition of blockchain technology, which is a distributed computing model that verifies and stores data through a cryptographic chain block structure, uses distributed nodes and consensus algorithms to generate and update data, and embeds automated scripting code (smart contracts) to program and manipulate data.[60] Blockchain adopts a time-stamp-based chain block structure, which basically operates as follows. First, the "miner" who obtains "bookkeeping rights" confirms the identity of the previous block by means of public key verification and private key signature and then reorganizes or decomposes the information passed by it and encrypts the reorganized or decomposed information with the private key of the block, so as to link the current block with the previous block to form a new blockchain. At

[59] Nakamoto [57].

[60] A related discussion of the definition of blockchain technology, see Antonopoulos [58], Swan [58], Underwood [59], Aste et al. [60], Pierro [61], Iansiti and Lakhani [62].

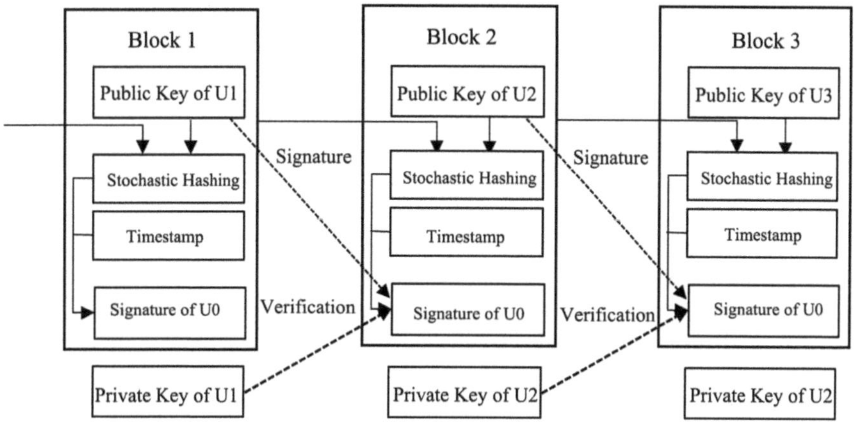

Fig. 4.7 Blockchain's Chain Structure Hash Principle[61]

the same time, when the block information is passed, each block "miner" needs to embed a random hash digital signature in the information and use it as the verification certificate of the information recipient. In addition, each block node records the data into its own block in chronological order, and each block is linked to each other in an orderly manner by a timestamp to form a blockchain, which is broadcasted in the block network as Proof of Existence of the block data. In this cycle, each block is connected in turn to form the longest master chain from the creation block to the current block, which completes the record of all data in the blockchain and traces and locates the data in this blockchain accordingly (Fig. 4.7).

Blockchain technology has the following four features: (1) Polycentric. Blockchain adopts an open-source polycentric and highly fault-tolerant distributed structure and verifies, records, stores, transmits, and updates data in multi-centralized and coordinated management. This distributed structure is able to ensure the consistency of distributed ledgers backed up across different nodes in the absence of central coordination and allows all members to participate in the management of collective data. As a result, damage to any node or loss of data does not affect the overall stability of the system and thus provides high security and reliability. Arvind Narayanan, an American computer scientist, and his research team believe that the polycentric or weakly centralized structural arrangement adopted by blockchain technology can maximize network-wide connectivity.[62] (2) Time-series backtracking. In fact, the time-series nature of the chain block structure greatly enhances the

[61] Yong and Feiyue [63].

[62] Narayanan argues that despite the decentralization potential of blockchain, at the current stage of blockchain development, if absolute decentralization is adopted, blockchain is unable to simultaneously satisfy the accuracy and efficiency of information recording, and is also prone to completely detach itself from the regulation of the mechanism. Therefore, the current blockchain application should still adopt a polycentric or weakly centralized structural arrangement with multiple nodes to verify, regulate and edit the data as a whole, so as to increase the efficiency of data exchange and the accuracy of cross-validation. See Narayanan [64].

traceability and verifiability of data. On the one hand, each piece of data in the blockchain is added with a time dimension, so it is possible to locate the position of the data in the block more accurately. On the other hand, the immutability and unforgeability of timestamps can fully guarantee the authenticity of data records. As Francesco Restuccia, a researcher in the Department of Computer Engineering at Northeastern University, says, blockchain can provide data protection for the system as a whole by adopting asymmetric encryption algorithms, providing a complete chain of evidence and a trusted way to trace data. Therefore, this feature can ensure a high degree of transparency and openness of data while effectively protecting the personal privacy of participants.[63] (3) Digital trust. Blockchain can build and maintain a complete, distributed, and tamper-evident ledger based on consensus mechanisms, asymmetric encryption, and a reliable database. Based on subgroup signatures and a distributed consensus mechanism combined with a hardware system, this non-interactive consensus mechanism allows individual nodes to endorse or guarantee the verification of transactions without the help of third-party institutions. For this reason, Benedikt Notheisen, a researcher at Karlsruhe Institute of Technology in Germany, points out that this feature of blockchain helps to realize value transfer and credit transfer.[64] (4) Intelligent programming. In addition to being able to be used for data storage and transmission, blockchain can also embed programmable operational code that can be executed by individual nodes accessing the network. Konstantinos Christidis, a researcher at the IBM Blockchain Institute in the United States, and others say that participants are able to program specific value restrictions, transaction pattern changes and updates, etc., into the blockchain and build smart contracts that are automatically triggered under specific conditions, thus automating transaction agreements or contracts.[65]

According to the scope of scalable nodes (open objects and scope) of blockchain, blockchain is generally divided into three application forms: Public Blockchain, Private Blockchain, and Consortium Blockchain. Firstly, a public blockchain, also known as Permissionless Blockchain, is a blockchain organization structure in which each node can freely access the blockchain according to the system specification and work based on a consensus mechanism. Public chains implement system maintenance mechanisms such as Proof of Work (PoW) or Proof of Stake (PoS). Therefore, public chains are considered to be more fully polycentric. Bitcoin, Ethereum, etc., are typical applications based on public chain architecture. However, because public chains require full node participation to process all transactions, the number of transactions they can process is very limited. This also reflects that blockchain will face

---

[63] Restuccia [65].

[64] In the traditional market transaction system, in order to reduce the trust cost caused by "limited rationality" and "opportunistic behavior," the main body of the transaction needs intermediary institutions such as trading platforms or arbitration institutions to regulate the transaction. On the other hand, blockchain establishes transactions on the basis of consensus and collaboration and expresses all the rules in the form of algorithms and programs in advance. As a result, participants can complete transactions without having to thoroughly understand each other's basic information. See Notheisen et al. [66].

[65] Christidis and Devetsikiotis [67].

the problem of inefficient processing when it is polycentric with more complete scalability. Second, a private chain, also known as a Permissioned Blockchain, is a monocentric network blockchain applied to data management and auditing within an organization. In a private chain, the eligibility of participating nodes is strictly restricted, and their write access is relatively more controllable. As a result, private chains can obtain faster transaction speed, lower transaction costs, better privacy protection, and higher security. When the National University of Singapore and Zhejiang University jointly launched BlockBench, an evaluation framework for private chains, they suggested that private chains can build more efficient data processing procedures and have better "Byzantine fault tolerance."[66] Finally, a federated chain is a blockchain composed of several institutions and maintained by these federated institutions. Consortium chains are also called Consortium Blockchains because they specify the read/write and bookkeeping rights of participating institutions according to the common rules of the consortium. The data in a consortium chain only allows different institutions in the system to read, write, and send transactions, so it is also regarded as partially polycentric. However, due to the relatively small number of nodes involved in consensus, consortium chains generally use consensus mechanisms such as Delegated Proof of Stake (DPoS), Practical Byzantine Fault Tolerant (PBFT), or Distributed Consistency Algorithm (RAFT).[67] Currently, the blockchain consortium R3CEV and Hyperledger, which is developed with the support of the Linux Foundation, are typical of the federated chain architecture.

On the basis of the above three application forms, Sidechain were born in order to enhance the scalability of the main chain and expand the innovation space of blockchain technology. In essence, sidechain are a solution to realize the transfer of digital assets across blockchains. In fact, the information isolation or blocking between many blockchains inevitably leads to the problem of multiple payments or value loss in information transfer. The essence of this problem lies in the "loss of frames" during the flow of data between different blockchains, and the value of the data chain is blocked due to the inability to communicate effectively between the information of each blockchain. The sidechain, on the premise of observing the "law of value conservation," forms a new blockchain by anchoring a node on the main blockchain and realizes the transfer of digital assets between multiple blockchains by using two-way anchoring and other mechanisms so as to break the "value island" caused by the precipitation of digital assets in their respective

[66] "Byzantine Fault Tolerance" refers to the fact that since it is impossible to achieve bilateral message consistency for message delivery over unreliable communication paths (where there is a possibility of message loss), distributed algorithms can be used to minimize the distortion of the message and to meet the specification of the message to be delivered. Therefore, Byzantine Fault Tolerance is of great value for multipoint networks that lack trust. See Dinh et al. [68].

[67] For example, Hyperledger establishes an authorization-distributed ledger on an invitation basis and exchanges value virtually and digitally. Most participants in Hyperledger reach a consensus based on a set of previously agreed-upon invariants and operate by balancing authorized and unauthorized nodes. See Harish Sukhwani, Jose Martinez and Xiaolin Chang, "Performance Modeling of PBFT Consensus Process for Permissioned Blockchain Network (Hyperledger Fabric)," paper presented to the conference on "2017 IEEE 36th Symposium on Reliable Distributed Systems (SRDS)," Hong Kong, Septmber 26–29, 2017, pp. 253–254.

blockchains. The problem of "value island" is caused by the deposit of digital assets in each blockchain.[68] At the same time, sidechain can also improve the privacy protection of information transfer between blockchains by establishing branch chains and providing a more secure way to upgrade the protocol for digital asset transfer. In this regard, Blockstream founders Adam Back, Matt Corallo, and others point out that the main purpose of sidechain is to avoid liquidity shortages and market volatility caused by indiscriminate digital currency issuance by enabling the transfer of different digital assets between multiple blockchains.[69] In this way, sidechain enable the construction of cryptocurrency financial ecosystems in a more integrated manner, allowing blockchain technology to be applied to asset types such as micro-payments, secure processing mechanisms, and property registration transactions.[70]

For the application levels of the blockchain technology paradigm, Melanie Swan, founder of the Blockchain Science Institute in the United States, divides it into three levels, namely, blockchain technology 1.0 for digital cryptocurrency represented by Bitcoin, blockchain technology 2.0 for digital cryptocurrency combined with smart contracts represented by Ether, and blockchain technology 3.0 for applications beyond cryptocurrency, finance and market.[71] First, Blockchain 1.0 refers to digital cryptocurrency and its payment system, whose main function is to realize the muti-centralization of digital currency issuance and means of payment, i.e., applying digital cryptocurrency in the process of transaction, settlement, and payment.[72] As of now, large financial institutions such as NYSE, ChiNext, Goldman Sachs, and Nasdaq have applied the achievements of blockchain 1.0 in the fields of cross-border transfer, remittance, and digital payment. Second, blockchain 2.0 refers to smart contracts. The core concept of this level of application lies in using blockchain as a programmable distributed credit infrastructure, thus using programs and algorithms to execute relevant transactions, i.e., smart contracts backed by algorithms and programs as a credit to handle various transactions.[73] At present, the typical

[68] The "law of conservation of value" mainly refers to the fact that the sidechains realize the synchronous updating of data between different ledgers, thus ensuring the avoidance of double payment or loss of value. Two-Way Peg is the technical basis of sidechains; its working principle is to lock the digital assets in the main chain temporarily and, at the same time, release the equivalent digital assets in the sidechains. When the equivalent digital asset is locked in the sidechains, the digital asset in the main chain can be released. See Christidis and Devetsikiotis [69], Dilleyet [70].

[69] Back [71].

[70] At present, the more representative sidechains are BTC Relay, based on digital cryptocurrencies; the Liquid open-source sidechains project developed by Blockstream, Lisk, based on non-digital cryptocurrencies; and domestic projects, such as Asch. See Jamie Redman, "BTC Relay The First Ethereum and Bitcoin Sidechains," LiveBIT, May 3, 2016, https://www.livebitcoinnews.com/btc-relay-the-first-ethereum-and-bitcoin-sidechains/; Samburaj Das, "Blockstream Announces Liquid—the First Sidechains for Bitcoin Exchanges," CCN, October 14, 2015, https://www.ccn.com/blockstream-announces-liquid-the-first-sidechains-for-bitcoin-exchanges/

[71] Swan [72].

[72] Swan [73].

[73] American computer scientist Nick Szabo first proposed the intelligent contract. The working mechanism of this technology is that when the data and information are incoming, the state of the resources in the contract resource collection will be updated, which in turn triggers the smart

representatives of Blockchain 2.0 are Ethereum, which is a public chain applied to the public, and Hyperledger, which is a federated chain applied to enterprises. The smart contracts based on Ethereum have been applied to various fields, such as electronic assets and consumption patterns.[74] Third, Blockchain 3.0 refers to blockchain technology that goes beyond economic areas such as currency, finance, and markets to non-economic areas such as government, healthcare, science, education, and the arts and reshapes the structure of human society and social relations with its advantages. *The Economist* mentioned in its October 2015 cover article *The Trust Machine* that blockchain-based ledger platforms have the potential to change the way people and businesses collaborate with each other, and that blockchain will see more experimental applications in finance, supply chain, trade, and other scenarios in the future.[75] This means that blockchain technology can be applied to all aspects of social life and thus become a global and universal technology embedded in the social infrastructure. As Efanov Dmitry and Roschin Pavel, researchers at the Moscow Institute of Engineering Physics in Russia, have argued, blockchain is, in fact, a universal technology, which in its 3.0 level of application, manifests itself as a digital smart society consisting of many horizontal cumulative elements.[76] At present, although Blockchain 3.0 is still at the stage of technical conception and has not formed systematic and large-scale applications, some market players have already conducted preliminary application trials in non-financial fields such as quantum-level management, automation of big data prediction tasks, distributed anti-censorship organization model and digital art authentication services (Fig. 4.8).[77]

It is worth noting that blockchain technology will also be closely integrated with the development of artificial intelligence technology in the future. From the current level of technological development, the significance of blockchain for AI is mainly reflected in the following three aspects.

---

contract to carry out state machine (Finite-State Machine, FSM) judgment. Suppose the triggering conditions of one or several actions in the state machine are satisfied. In that case, the state machine will select the prior contract action according to the preset information and execute the contract automatically, i.e., the digitized promises will be executed automatically when the triggering conditions are satisfied. Of course, the smart contract itself is only a transaction processing system and does not modify the content of the contract itself. See Szabo [74].

[74] Koulu [75].

[75] "The Trust Machine: The Technology behind Bitcoin could Transform how the Economy Work," *The Economist*, October 31, 2015, http://www.economist.com/news/leaders/21677198-technology-behind-bitcoin-could-transform-how-economy-works-trust-machine

[76] Dmitry and Pavel [76].

[77] For example, Monegraph, a new digital art and media platform, provides services such as authentication, transmission, and trading of digital artworks; Namecoin provides a distributed cryptographic solution for storing and transferring matching secret keys and property ownership; and Factom applies blockchain technology to the protection and verification of data such as public records and business documents. See Harry Kalodner, Miles Carlsten, Paul Ellenbogen, Joseph Bonneau and Arvind Narayanan, "An Empirical Study of Namecoin and Lessons for Decentralized Namespace Design," *WEIS*, 2015, pp. 1–2; Rachel Dwyer, "Does Digital Culture Want to be Free? How blockchains are transforming the economy of cultural goods," *Academia*, January 3, 2019, http://www.academia.edu/33838249/Does_digital_culture_want_to_be_free_How_blockchains_are_transforming_the_economy_of_cultural_goods.

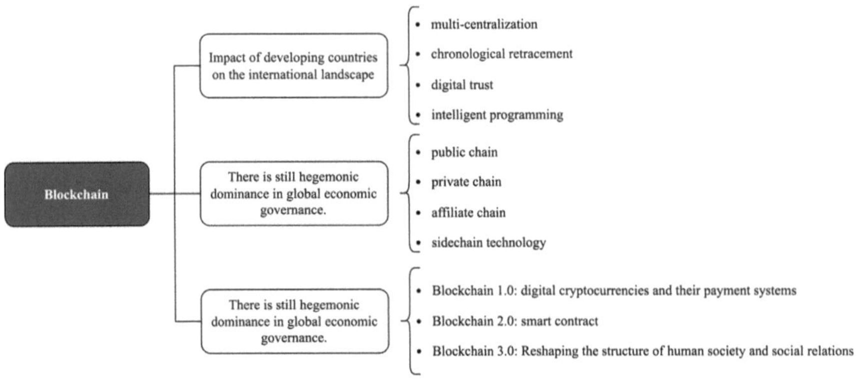

Fig. 4.8 The Characteristics, Application Patterns and Levels of Blockchain

First, blockchain can further enhance the effectiveness of AI by achieving multi-centered connectivity. Although AI can provide new labor increments for productivity, AI cannot achieve the equalization of development factors. This is due to the fact that the data generated by a large number of smart body applications remain monopolized by intermediary platforms. This data-centric development model then leads to under-exploitation and under-utilization of productivity.[78] The intervention of blockchain can build an information-sharing platform based on a distributed system for the development of AI and realize the data clustering and evolution based on this platform in combination with digital credit. This system cluster and platform optimization can not only promote the sharing of data but also realize the centralized management of data.

Secondly, blockchain can lubricate social relationships in the era of artificial intelligence by building digital trust. The most typical feature of the intelligent society is the emergence of a large number of intelligent agents as new actors, and the emergence of intelligences will give rise to the formation of massive data and complex and changeable social relationships accordingly.[79] Blockchain technology can take over the demand of smart society for data privacy and security protection. On the one hand, the intervention of blockchain can provide a high level of privacy protection for data, i.e., through asymmetric encryption, public and private keys, and other designs to achieve data encryption preservation and authorized use of data.[80] On the other hand, the peer-to-peer data recording method and distributed storage method adopted by blockchain can prevent data forgery to the greatest extent and thus ensure the security and reliability of artificial intelligence by improving the transparency and accuracy of data.[81]

[78] Qiqi [77].

[79] Daugherty and Wilson [78].

[80] Kshetri [79].

[81] Vigna and Casey [80].

Third, blockchain can promote "individual-centered" data aggregation and give individuals more say in data usage, based on data sharing and security. Through inductive, retroactive, and deductive interactions, blockchain can facilitate the formation of social credit relationship networks. This can activate more application scenarios centered around individuals. In turn, this provides artificial intelligence with a self-trusted network. The combination of the two technologies can effectively create more value for the application of artificial intelligence in a self-trusted network. Obviously, the combination of the two technologies can effectively handle the large amount of data generated between these actors and can also better realize the intelligence of these social relationships based on digital credit, thus improving the overall operational efficiency of the intelligent society.

It can be seen that the combination of blockchain and artificial intelligence not only helps to improve each other's application space but also provides a better means of technical supervision and accountability. In fact, in the long run, the future development of social productivity needs to be centered on artificial intelligence, while the adjustment of production relations needs to be centered on blockchain technology. This is due to the fact that the biggest role of AI is to liberate labor, i.e., to replace part of the labor force's work through computer simulation, so what AI achieves is a revolution in productivity.[82] Blockchain, on the other hand, embodies a function of doing value work again and reorganizing people, so it is more of a revolution of production relations. In fact, the goal of global governance is to achieve balanced development and global justice, which should be based on equal resources and equal capabilities.[83] The combination of blockchain and artificial intelligence can create more opportunities to achieve equality of ability by improving productivity on the one hand and achieving a more equitable distribution of resources by optimizing production relations on the other. Therefore, under the resonance of blockchain and artificial intelligence, we can realize a social order in which productivity and production relations progress together and thus promote the realization of global governance goals.

Of course, we must distinguish blockchain technology from the digital cryptocurrency represented by Bitcoin. In fact, many experts and scholars have questioned digital currencies such as Bitcoin. Nobel laureates in economics, such as Robert Shiller, Paul Krugman, Richard Thaler, Joseph Stiglitz, and Eugene Fama have argued that Bitcoin is more of a "bubble" due to its nature and market factors.[84] American economist Nouriel Roubini has even argued that Bitcoin is the "mother of all bubbles".[85] Meanwhile, Professor Rainer Böhme of the University of Münster, Professor Nicolas Christin of Carnegie Mellon University, Professor Benjamin Edelman of Harvard Business School, and Assistant Professor Tyler Moore of the University of Tulsa in Oklahoma, USA, have been writing about Bitcoin in the field

[82] Makridakis [81].

[83] Qiqi [82].

[84] Fadilpašić [83].

[85] Rooney [84]. For other related discussions on whether Bitcoin is a bubble, see Fry et al. [85], Pavel Ciaian, Rajcaniova [86], Bjerg [87].

of economics. Tyler Moore has published an article in *Journal of Economic Perspectives*, one of the most influential and top-tier journals in the field of economics, providing a systematic analysis of the market risk, counterparty risk, operational risk, and privacy-related risks associated with Bitcoin.[86] In addition, Yao Qian, former director of the Institute of Digital Currency at the Central Bank of China, discusses digital currencies from the perspectives of value stability, public economics, and transaction cost theory and argues that private digital currencies such as Bitcoin cannot be a true currency.[87]

Specifically, Bitcoin has the following main problems. First, although Bitcoin can solve the credit problem of transactions to a certain extent and play the role of a medium of exchange in certain scenarios, Bitcoin, as an algorithmic workload or proof of ownership, provides only a rough measure of monetary metrics, lacks the regulatory mechanism to maintain the actual monetary exchange value, and does not have the value scale of monetary function. For example, Pavel Ciaian, a principal researcher at the European Commission's Joint Research Centre, argues that because Bitcoin adopts a constant scale issuance mechanism, it cannot be adjusted to real economic developments in a timely and effective manner. As a result, Bitcoin is unable to produce long-term stable optimization effects on the real economy and financial system.[88] Second, the current Bitcoin ecosystem has determined that the speculative nature of Bitcoin far exceeds its real-world value, and some speculators even speculate on Bitcoin by building so-called "coin circles."[89] Third, Bitcoin's issuance structure, operational transparency, digital wallet and exchange security are all extremely risky, and therefore, with the manipulation of capital forces, Bitcoin has become a danger to economic order and social stability. Danton Bryans, a researcher at Indiana University Morrill College of Law, argues that Bitcoin's own flaws and lack of regulation have often made it available for criminal activity, making it a dark web medium of exchange, a tool for financial fraud and market speculation.[90] For example, the earliest virtual currency exchange, BTC-e, is suspected of using Bitcoin to launder more than $4 billion for criminal organizations, while the largest dark web black market, AlphaBay, uses tokens such as Bitcoin as its primary payment instrument and as an "accomplice" to money laundering. "AlphaBay" is the largest dark web black market. It can be seen that although digital currencies such as Bitcoin are the most mature application of blockchain, digital cryptocurrencies such as Bitcoin are only the embodiment of blockchain 1.0 technology bearers, and there are many uncertainties and hidden risks of their own.

The revolutionary effect of blockchain technology on production relations is mainly manifested in the social revolution to overcome the problem of collective

[86] Böhme [88].
[87] Qian [89].
[88] Ciaian and Rajcaniova [90].
[89] Elbeck and Baek [91].
[90] Bryans [92].

action, the value revolution to break traditional intermediary pricing, and the governance revolution to achieve maximum social participation. Therefore, the application of this technology has the following impacts on global economic governance: First, blockchain can create good technical conditions for the construction of self-trust ecology, promote the integration of multiple mechanisms oriented to collective interests, and accelerate the transformation of global economic governance from centralized to polycentric. In fact, information asymmetry in the market mainly occurs in the issues of adverse selection before the transaction and moral hazard after the transaction. Blockchain is able to build a polycentric and trustworthy system through digital cryptography and distributed consensus algorithms. This feature enables global economic governance actors to use a polycentric model to achieve self-certification of each node of the network without trusting a single node and to generate digital trust through "Coded Trust".[91] Brett Scott, a senior researcher at the UK Financial Innovation Lab, believes that in an ecology of digital trust, global economic governance institutions can bring more actors into the governance system at a lower institutional cost. And under this new trust mechanism, actors do not need to know basic information about each other's transactions in order to engage in trustworthy value exchanges.[92] This means that blockchain can rely on the trust and consensus created by digital law to enable individuals to transfer value without the need for third-party financial intermediaries. This helps break the shackles of institutional inertia and vested interests in multi-party cooperation and creates a good ecology for bilateral trust and multilateral cooperation to form. At the same time, blockchain can reduce the emergence of the "Lemon Market" and effectively enhance the stability and durability of cooperation by verifying the information of economic activities through high-trust nodes and realizing the transfer of assets based on the principle of global consistency.[93] Sarah Manski, an American globalization scholar, argues that blockchain can greatly increase the importance of social capital so that the "shareable value" formed through collaboration will replace the exchange value in the global capitalist market.[94] Radmehr Monfared, a researcher at the Centre for Innovative Manufacturing in Intelligent Automation at Loughborough University in the UK, says that blockchain can help reduce divergences in global economic governance in terms of collaboration and make it more responsive to the need for public goods, thereby facilitating a shift from monopolistic and resource-advantaged

---

[91] "Coded Trust" is a derivative concept of the Programmable Economy, which refers to trust mechanisms that are "editable and controllable" by incorporating entities into an all-encompassing programming structure in order to autonomously support and manage the generation, production, and consumption of goods and services, as well as to support the anonymous, encrypted exchange of multiple values (monetary and non-monetary) in different scenarios. See Casey and Vigna [93], Furlonger and Valdes [94].

[92] Scott [95].

[93] "Lemon Market" refers to a free market in which, under conditions of information asymmetry, sellers will have more information than buyers, leading to the expulsion of high-quality products from low-quality ones. See Akerlof [96].

[94] Manski [97].

global economic governance to an open and service-oriented one.[95] In addition, the distributed data storage structure of blockchain can realize the diversion of data and information and also avoid the problem of disruption of the overall structure due to the damage or absence of a link in the centralized structure, thus reducing the transaction cost of the original centralized structure system. In this regard, Richard Adams, a researcher at Cranfield University in the United Kingdom, pointed out that blockchain can guarantee information security while ensuring the high efficiency and low cost of information system operation, thus promoting the networking and globalization of economic information and realizing the transformation of information Internet to value Internet.

Secondly, blockchain can help clarify and maintain the relationship among various actors in global economic governance and provide institutional guarantee for the interaction practice between global economic governance and domestic economic governance of each country. In fact, it is difficult to avoid the phenomenon of "free-riding" in collective action in global economic governance, and the fragmented relationship exacerbates the cost of communication among different actors. Therefore, it is difficult for the global economic governance system to break the shackles of this cooperation dilemma. The intervention of blockchain technology will build a logic of mutual governance in global economic governance.[96] First, blockchain can store all data information in the form of a distributed data system and ensure the traceability and accountability of the system as a whole with its open source technology and data sharing, which in turn helps to realize the improvement of information symmetry and the reduction of trust cost. According to William Nikolakis, associate professor at the University of British Columbia, blockchain will help to handle the huge amount of data information generated by the development of global value chains and can guarantee the symmetry of such data information in terms of "provision, verification, and enforcement."[97] This means that blockchain can provide more convenient technical support for information communication and policy coordination among various actors. This will, to a certain extent, strengthen the interaction between the relatively independent regulatory systems and rules, thus laying a good communication foundation for the formation of a highly integrated and comprehensive global economic governance system.[98] At the same time, blockchain can be used to record all kinds of structured or unstructured data information and to process, quantify and evaluate these data by establishing big data algorithm programs and distributed computing methods, so as to maximize the use of all existing data information. This feature can help monitor actors more effectively and provide more reasonable indicators

---

[95] Abeyratne and Monfared [98].

[96] As Hayek said, "Order is a state of things: a state in which a multiplicity of factors are interconnected in such a way that we can derive correct expectations for other parts from familiar parts of space or time, or at least make it possible for us to derive correct expectations." Thus, sufficient conditions for mutual assistance between subjects are only present when a more stable relationship of trust is formed. Hayek [99].

[97] Nikolakis et al. [100].

[98] Lemieux [101].

for "selective incentives" in global economic governance, thus reducing compliance problems caused by regulatory fragmentation and bridging regulatory conflicts and loopholes among them. In addition, although the emergence of the Internet has made online reputation evaluation systems and autonomous selection mechanisms an important channel for forming trust between parties to transactions, there is still a risk of manipulation and tampering with these systems and mechanisms.[99] Blockchain, on the other hand, can not only greatly reduce data information asymmetry, but its unique consensus mechanism can also help strengthen the verification of the authenticity and integrity of transaction contents. According to Melanie Swan, blockchain can more objectively record the extent and deviation of each actor in the implementation of specific rules and fully utilize the utility of data in governance as well as provide a new model for consensus building.[100] Mark Coeckelbergh, a professor at the University of Vienna, Austria, believes that blockchain can maximize the neutrality and openness of data, which will significantly improve the legitimacy and credibility of transactions.[101]

Thirdly, blockchain can create a more free and open system collaboration mechanism through clever technical design and economic incentives, which can better meet the needs of multilateral collaboration under the deep development of economic globalization. Firstly, blockchain can effectively improve the convenience of international trade documents and agreements through its transparency and immutability and minimize the friction in value exchange by means of asymmetric encryption to reduce the cost of reaching consensus among participating parties. Marco Iansiti, a professor at Harvard Business School, believes that the distributed, peer-to-peer transmission, unmodifiable data records, and transparent anonymity of blockchain can reduce the cost of connection among actors and build an open relationship among the connected parties, thus promoting the formation of a new language of social cooperation to unlock new economic values.[102] Second, blockchain can improve the flow and trustworthiness of data and information, enabling multinational enterprises to share data, trade goods, and finance directly through blockchain and completely connect upstream and downstream enterprises in the supply chain to form industry-based collaboration groups or collaboration blocks. More importantly, blockchain can break the monopoly of financial intermediaries on information by building a new trust mechanism, thus reducing the transaction risk and cost of each actor while breaking any national, regional, or institutional restrictions on data information. This technical characteristic can not only improve the efficiency of using financial capital but also significantly improve the quality and added value of products, thus promoting the multi-faceted derivation of the global value chain.[103] Third, blockchain can realize the automatic processing of data through algorithm and program design and build the underlying protocols of the payment and settlement system, thus optimizing the

99 Luca [102].

100 Swan [103].

101 Coeckelbergh and Reijers [104].

102 Iansiti and Lakhani [62].

103 Kshetri [105].

payment system in terms of credit system construction, simplifying the transaction process, and improving the security of transactions, as well as enhancing the risk control level of operation links. At present, some commercial banks in China have already started to try to use blockchain technology for cross-border transfers. For example, China Merchants Bank, together with Wing Lung Bank and Wing Lung Shenzhen Branch, has successfully used blockchain technology to realize cross-border RMB remittances.[104] In addition, in combination with the asset tagging and identification technology of the Internet of Things, blockchain can also be applied to the records of the whole process of production, transaction, and transportation of offline products. It can guarantee the quality of products in the form of digital verification, promote the electronic unification of the three flows of logistics, information flow, and capital flow, and realize the flexible management of the supply chain and product traceability and other functions. For example, IBM has embedded blockchain technology into its large cloud infrastructure and applied this to track high-value goods in the supply chain.[105]

Overall, although there are few blockchain projects that have actually landed and generated some social benefits, and blockchain itself is still in the preliminary stage of technology development, we still need to see that blockchain can realize the construction of polycentric architecture and distributed transactions by using technologies such as cryptographic chain block structure, distributed nodes, and consensus algorithms. The application of this technology in global economic governance can maximize the plurality of participants, clarify and maintain the relationship between various actors in global economic governance, and create a new type of free and open system of collaboration mechanism. Therefore, the application of blockchain can improve the synergy of global economic governance mechanisms and promote the reconstruction of international trade and the global industrial chain. On this basis, blockchain will help build a fairer and more inclusive global economic governance model and promote structural reform of the global economic order.

4.4 The Idea of a Supersovereign Currency: The Limits of SDR and the Possibility of E-SDR

As the core of global economic governance, the reform and transformation of the international monetary system have always been an essential issue for the participation of all countries. In this regard, whether from the perspective of the history of the international monetary system or from realistic development needs, the creation of a super sovereign international currency is an ideal solution, because it is independent of a country's economic interests and situation and can effectively coordinate the

[104] "Blockchain helps China Merchants Bank complete the world's first 'most perfect cross-border payment,'" China Merchants Bank, December 21, 2018, http://www.cmbchina.com/cmbinfo/news/newsinfo.aspx?guid=8a3d1509-dc5d-45b3-a7c2-80f9b5c423d8.

[105] Jackson [106].

interests of all parties. At present, although the attempt represented by SDR is in line with the development direction of the international monetary system, SDR cannot fully realize its responsibility as a super-sovereign currency due to its defects such as lack of credit base, narrow scope of use, and insufficient consensus mechanism, as well as interference from many external factors. With blockchain technology as its core, the E-SDR can provide a new perspective for constructing a super sovereign currency system.

The international economics community has expected to create a super sovereign monetary system free from manipulation by major powers. Henry White, the former Assistant Secretary of the Treasury of the U.S., was the first to propose the "International Stability Fund Plan," or the so-called "White Solution." The proposal calls for creating an International Monetary Stabilization Fund (IMSF), with voting weight in the IMF determined by the shares paid by member countries. The organization would issue an international currency called the Unita, which would be convertible into gold and transferable between member countries.[106] The British economist John Keynes proposed the so-called "Keynesian Solution". The program advocates the establishment of a global central bank "international clearing union". It advocates the issuance of a "Bancor" as the international monetary unit and national currencies with the "Bancor".[107] The Bancor is a fixed but adjustable exchange rate. The "Bancor" and the "Unita" are already super sovereign reserve currency concepts with a certain degree of independence decoupled from the sovereign state.

Many scholars have further proposed more creative ideas based on these two conceptions and related practices. Based on Keynes, Triffin proposed a "reserve concentration scheme" that would internationalize national foreign exchange reserves by creating additional import and export funds and adjusting the supply of deposits (reserves) to meet the liquidity needs of the world economy. The plan also envisages the creation of an international currency to reduce the pressure on the U.S. dollar as the main reserve currency.[108] The British economist Friedrich Hayek proposed the theory of denationalization and monetary competition, whereby the

---

[106] The Unita is a unit of account rather than a currency, and the currency remain in circulation are still sovereign currencies such as the US dollar. At the same time, the Unita is issued on a fund based basis, and once a sovereign country completes its subscription, its share is relatively fixed. Therefore, this mechanism still lacks a substantial withdrawal channel and currency derivation mechanism, making it difficult to adapt to global economic and trade growth automatically. See Xiaoling and Ge [107].

[107] The "Bancor" is a hypothetical departure from the gold standard to return to the use value of money. On the one hand, "Bancor" refers to 30 kinds of representative commodities composed of the currency basket as the value of support. On the other hand, Keynes proposed that all international trade be denominated in the "Bancor" and that surplus countries be forced to deposit surpluses in allied accounts and provide overdrafts to deficit countries, placing greater emphasis on exchange rate flexibility. Thus, the "Bancor" has more of a super-sovereign currency connotation than the "Unita." See Gardner [108].

[108] The program proposed by Triffin is actually a restructuring of the International Monetary Fund, and the vision for the construction of a super-sovereign currency is limited to the solution of the problems of global liquidity and insufficient growth of currency reserves. See Robert Triffin, "Gold and the Dollar Crisis," *Challenge*, Vol. 9, No. 2, 1960, pp. 40–43.

state would issue monopoly currencies instead of competitive ones issued by private institutions and by selecting winners and losers. The theory of denationalization and monetary competition is to replace monopoly currencies with private institutions and select stable and reputable currencies.[109] The American economist Robert Mundell proposed the idea of Optimal Currency Areas (OCAs), which is to establish optimal currency areas in countries and regions with close economic interactions through fixed exchange rates or unified currencies and to offset the impact of exchange rate uncertainty on the economy.[110] Richard Cooper, a professor at Harvard University, proposed the idea of establishing a single currency system, that is, through the establishment of a global central bank and its control of the issuance of world currency, a unified monetary system and a unified monetary policy will be gradually established in the world.[111] Italian economist Pietro Alessandrini and others proposed the establishment of a supranational bank currency (SBM) in the institutional environment of a clearing union.[112]

To this end, the International Monetary Fund developed and implemented the Special Drawing Rights (SDR) program in 1969. The SDR is a reserve asset and unit of account created to meet international liquidity needs and supplement the international reserve currency. (1) As a value yardstick, the SDR can serve as a reference standard for sovereign currency exchange rates, i.e., the SDR serves as a relatively stable currency basket and thus provides a reference basis for international exchange rates. (2) As a value reserve instrument, the SDR is equivalent to other

[109] According to Hayek, the state monopoly of issuing money exists for more disadvantages than advantages because (1) the government monopolizes the right to issue money out of self-interest, i.e., the government takes a mint tax from the issuance of money and circulates bad money instead of good money; (2) when the government obtains the power to take care of certain social classes or groups of people partially, the government forces these classes or groups to exercise this power in order to obtain sufficient support for it; and when paper money is put under political control instead of metallic money, the government becomes the typical representative of this power; (3) the monopoly of money supports the expansion of government power, and the government thus continues to over-issue money, which in turn exacerbates currency devaluation and inflation. See Hayek [109].

[110] Mundell believes that if the countries in a region want to realize the full integration of the currency, they can establish a common currency in a specific region or implement a rigid system of fixed exchange rates for national currencies, the foreign exchange rate of the unified implementation of the floating system. To this end, Mundell put forward the "financial stability of the three islands" idea, that is, Europe, the Americas, and Asia, to form their monetary unions, and then the three parts to form a global monetary union. Based on Mundell's theory, several economists from different aspects of the super-sovereign currency discussion, for example, McKinnon that the degree of openness of the economy as the best currency area of the judgment standard, and proposed in some of the close trade relations between the economic open areas to form a relatively closed common currency area, and the implementation of a fixed exchange rate in this region, the implementation of flexible exchange rate system; Princeton University Professor of Economics Peter Kenen that the product diversification of countries and regions is the ideal object of the construction of the optimal currency area; University of Northern Colorado Professor of Economics James Ingram will be introduced into the study of the optimal currency area of financial integration. See Mundell [110], Mckinnon [111], Kenen [112], Ingram [113].

[111] Cooper [114].

[112] Alessandrini and Fratianni [115].

foreign exchange reserves. It can be exchanged with freely usable currencies to become a reserve asset of IMF member countries. (3) As a medium of exchange, SDRs can be used for settlement between the IMF and member countries and among member countries. In particular, when a country holding SDR with a balance of payments deficit, it can use its quota to exchange foreign currency with other member countries to pay off the balance of payments deficit or repay the IMF's loans and interest.[113] The value of the SDR is determined by the basket currency exchange rate, which can offset the impact of exchange rate fluctuations on a single currency to a certain extent. At the same time, the IMF allocates SDR shares mainly according to the proportion of the amount paid by member countries, and the IMF will adjust the weight of the currency basket according to the changes in economic development.[114] For this reason, many scholars propose to make the SDR move toward a world currency by giving full play to its role. Stiglitz pointed out that a critical path to reform the dollar-centered international monetary system lies in increasing the role of the SDR and therefore needs to promote a wider distribution of the SDR and expand its scope of application. José Ocampo, a professor of economics at Columbia University, advocates transforming the SDR into the dominant international reserve asset to maintain the overall stability of the international monetary system.[115] Zhou Xiaochuan, former governor of the People's Bank of China, believes that SDR, as a practice of super sovereign currencies, not only have the advantages of a stable value, ease of use, and low cost of acquisition but also can overcome the inherent risks of sovereign credit currencies and regulate global liquidity more effectively.[116]

However, the SDR also has many flaws, and therefore there are many arguments that the SDR needs to have the ability to become an international currency. First, the SDR is still based on the credit currencies of its member countries and contains only five sovereign credit currencies (in which the U.S. dollar still dominates). This means that the SDR cannot guarantee its ultimate value and fails to change the international monetary system's reliance on the sovereign currencies of large countries as reserve assets.[117] At the same time, the SDR lacks the credit base to maintain currency stability. In this regard, Indian economist Swaminathan Aiyar pointed out that the current international mutual trust is still insufficient, so there is no global social basis for transferring "monetary sovereignty," countries are still more cautious about

---

[113] IMF, "International Monetary Fund: Special Drawing Rights," April 19, 2018, https://www.imf.org/en/About/Factsheets/Sheets/2016/08/01/14/51/Special-Drawing-Right-SDR.

[114] Since October 1, 2016, the SDR currency basket has been expanded accordingly to five currencies: the US dollar, the Euro, the Chinese Yuan, the Japanese Yen, and the British Pound, with weights of 41.73%, 30.93%, 10.92%, 8.33%, and 8.09%, respectively. IMF, "International Monetary Fund: Special Drawing Rights," April 19, 2018, https://www.imf.org/en/About/Factsheets/Sheets/2016/08/01/14/51/Special-Drawing-Right-SDR

[115] Ocampo [116].

[116] Xiaochuan [117].

[117] Aiyar [118].

authorizing the IMF to issue SDRs.[118] In addition, the number of SDRs issued is still relatively limited (SDRs currently account for only about 5% of global reserve assets). SDRs cannot be used for official-private and inter-private settlements and thus cannot be used directly in international financial activities. Therefore, given its current size and scope, the SDR is not effective in meeting the needs of international trade settlements and international reserves.[119] Eichengreen argues that the SDR's scope of use limits it from being used in regulating foreign exchange markets and by market participants in general.[120]

From the above-mentioned economists' conception of a super sovereign currency and the practice of SDR, a super sovereign currency can, to a certain extent, overcome the inherent risks of a sovereign credit currency and provide the possibility of regulating global liquidity. However, given the current state of the international political economy and the shortcomings of the SDR, it is still necessary to continue to explore the construction path of a super sovereign currency system.[121] To this end, it can be believed that we can build a set of E-SDR super-sovereign digital currency system based on blockchain data encryption, distributed consensus, and timestamp technology, with the basic structure of "digital currency system, digital financial account system, and digital identity verification system", and use it to promote the change of global economic governance system.

The specific components of the E-SDR are as follows: (1) The digital currency system, as the underlying structure of the E-SDR, is the basic unit of account for each participating node to transfer funds. The use of digital currency is characterized by data temporality, immutability, and forgery and can realize peer-to-peer transactions under a self-trusting system ecology. Digital currency can adopt the mechanism of direct public issuance by the central bank or the mechanism of joint issuance by the central banks and commercial banks.[122] Yao Qian has suggested that a binary model of "central bank-commercial bank" can be used to issue digital currency, i.e.,

---

[118] According to Aiyar, since the SDR is a "derivative" issued based on member countries' currencies, it does not meet the conditions for currency creation, and the IMF itself, as an international financial institution, does not meet the basic requirements for currency issuance. See Aiyar [119].

[119] Aiyar [120].

[120] Eichengreen [122].

[121] Stei [123].

[122] Currently, the issuance methods of digital currencies mainly include legal tender digital currencies, digital currencies anchored to fiat currencies or assets, digital currencies based on crowdfunding project funds, and digital currencies based on algorithms. Among them, the issuance mode of legal tender digital currencies is mainly categorized into three types: (1) the central bank issues digital currencies directly to the public, and individuals, enterprises, and financial institutions simultaneously intervene in the central bank's system. This issuance model is the most direct and thorough and can ensure that the central bank obtains the greatest initiative, but the lack of a transition phase and transition medium is likely to confuse the current financial system; (2) based on retaining the existing currency issuance, the central bank cooperates with commercial banks to jointly issue digital currency. This model can achieve a smooth transition, but it will weaken the central bank's right to control; (3) continuing the current decentralized issuance model for digital currencies, but in the end, the central bank will unify the bookkeeping model and conversion mechanism. This model is conducive to the competition of digital currencies but will likely confuse the digital currency

the central bank is responsible for issuing and verifying digital currency monitoring, and commercial banks are responsible for providing circulation services.[123] (2) The digital financial account system, as the middle layer of E-SDR, covers four major account systems, namely, the central bank payment system, the financial account system under the central bank payment system, the accounting system opened by financial institutions for legal persons and the accounting system opened by financial institutions for natural persons. This architecture layer can adopt the "law of clustering". Each subject's account is categorized into a unified node and then directly linked to various subjects, thus realizing the interconnection of nodes in the complex network system.[124] (3) The digital identity verification system is the top-level architecture of E-SDR, and the mapping set of the authorized middle-level architecture (digital financial accounts are the sub-accounts of this layer). Each sub-account is granted a unique legal person identifier (institutional code) or natural person identifier (biometric code). The layered architecture must authorize the transaction process to verify the signature code of the subject of the transaction before the transfer of funds can be completed.

The composition of this mechanism of E-SDR evolved from the three-layer structure of cash, financial account, and account encryption of the original credit currency system. However, E-SDR differs significantly from the traditional sovereign credit currency system and the digitization of money. This is due to the fact that the construction of the E-SDR, in addition to covering the cash layer of the original monetary system, necessitates the creation of digital financial accounts and digital authentication methods to accompany the digital currency. Only when these three layers are systematically built, and a completely digital currency ecosystem is formed can E-SDR be genuinely realized. E-SDR treats the financial system as a complex network system in which financial institutions are the nodes of the complex network system, and currency is the medium of data and information circulation between different nodes. In addition, when E-SDR is realized, the financial, fiscal, tax, and social security systems can be embedded in it, forming a new social governance framework. As Wessel Reijers, a researcher at the Robert Schuman Research Center at the European Union University, says, the economic application of blockchain is only its most intuitive and superficial form of action; the trust transformation mechanism behind this technology will restructure social relations and form a technical, social norm based

---

market. For related studies on the issuance mechanism of digital currencies, see Giaglis and Kypriotaki [124], Ali et al. [125], Bech and Garratt [126], Fung and Halaburda [127], Masciandaro [128].

123 Qian and Tang [129].

124 The "clustering principle" refers to the use of distributed multi-node collaborative processing to categorize the different types of property owned by legal persons and natural persons and unify them with the corresponding accounts of legal persons or natural persons to ensure that the mapping relationship between the digital financial account system and the subjects tends to be a complete one-to-one mapping, and thus to realize the sharing of information and value in a more efficient manner. See Fanning and Centers [130].

on digital trust.[125] Therefore, the E-SDR embedded in blockchain technology also has the potential to become the underlying framework for social governance.

Specifically, the E-SDR has the following characteristics and roles. First, as a way to explore a super sovereign currency, the E-SDR is a unified, complete, and independent monetary system. Therefore, the E-SDR can serve as a relatively neutral clearing, settlement, and transaction method. In essence, a currency is a materialization of a publicly accepted claim to future value, i.e., a currency is a social consensus formed under rules based on unanimous consent.[126] In the commodity and credit money stages, money anchoring is based on the scarcity of commodities and the credit backing of sovereign states, respectively. E-SDR realizes the reshaping of the monetary credit system through technical and institutionalized arrangements. On the one hand, based on the distributed bookkeeping technology of blockchain, E-SDR can provide a more effective and convenient channel for the use and circulation of currency.[127] This is because the application of big data, cloud computing, and other technologies has built a bookkeeping and clearing mechanism for various social transactions. Blockchain can promote direct peer-to-peer transactions on top of this mechanism by building a distributed ledger, thus improving the liquidity of the entire economic system and reducing the credit and transaction costs of currency use. According to American economist Reuben Grinberg, distributed ledgers can not only achieve a certain degree of self-confidence in the value of money and consensus in its use but also improve the interface between goods and money.[128] On the other hand, applying blockchain and technologies such as big data and cloud computing can structure the analysis of the connection between various commodities and currencies. Based on this real-time data analysis, we can build a "basket price index" that can reflect the overall value of the commodity, and the E-SDR will be based on this index and adjusted dynamically according to the fluctuations of the index.[129] This means that the E-SDR can be anchored to society's overall production capacity and the value of commodities. Of course, the anchoring of E-SDR with this "basket index" needs to be based on the full completion of social credit, i.e., E-SDR realizes its social nature with the backing of complete digital credit.[130] The digital trust, distributed bookkeeping, and wide anchoring of E-SDR not only help to reduce the

[125] Reijers and Coeckelbergh [131].

[126] Yeomans [132].

[127] The credit cost of money mainly refers to the need for the central bank to provide specific credit certificates for issuing money, and its transaction cost refers to the need for market players to pay transaction fees for the financial intermediaries they rely on. See Doguet [133].

[128] Grinberg [134].

[129] Chen Xiangguang, a professor at the School of Economics of Renmin University of China, proposes that the "basket of energy, the basket of gold reserves of sovereign states, and the composite of various types of currency issuances in the Special Drawing Rights" can be used as the anchor of the future legal tender digital currency. See Xiangguang and Zeqing [135].

[130] Due to the high volatility and complexity of the currency multiplier influencing factor of this "basket index," the accuracy of the credit creation capacity of the E-SDR and its regulatory capacity will also be affected to a certain extent. See Lei and Chengguo [136].

frictional cost of the original value exchange but also effectively guarantee the health of its own issuance and smooth flow.

Second, the embedded digital account layer enables E-SDR to collect and analyze transaction information among all nodes and terminals, thus enabling market factors such as global economic growth and inflation levels to be more fully incorporated into the money supply. According to Burmai and other scholars, with the improvement of variables and financial information loading, the application of blockchain in the digital currency layer can lay a sufficient foundation for "automatic monetary policy,"[131] which is based entirely on nominal data. E-SDR can create a direct contact channel between the central bank and the public, and digitally transform from cash clearing to book-entry clearing. As a result, E-SDR can be more closely integrated with a series of economic activities such as goods and services, helping policymakers to grasp overall market activities more flexibly and accurately and providing more effective monetary and price tools for macroeconomic regulation, thus keeping the quantity of money within a reasonable range to meet the balance of supply and demand in the market. At the same time, E-SDR can build a training model for policy expectation adjustment with the help of artificial intelligence and big data, and other technologies to predict the reasonable amount of digital currency to be placed and the forward-looking conditions for returning interest rate adjustment at a certain point in the future.[132] Rahman believes that the intervention of blockchain in digital currency can respond to the market demand for money in a timely and adequate manner, and under the appropriate regulation of the government, this feature can help improve the utility of monetary policy and achieve a more balanced "optimal monetary policy".[133] Therefore, this digital correlation feature of E-SDR can provide a more effective decision-making tool for monetary and fiscal policies, which will, to a certain extent, promote the international monetary system to match the level of global economic development better.[134]

Third, E-SDR will improve the payment and settlement system regarding the transaction process, transaction security, and system traceability and build a public platform connecting all nodes in the system. Under the traditional financial system, most payment behaviors are merely the flow of monetary data in different accounts, while vital information such as the reason for payment, the source of funds, and the path of funds circulation are stored in different payment environments and systems. As Shinto Teramoto, a professor at Kyushu University, says, the most central problem in traditional financial system arrangements is information asymmetry due to limited cognitive capacity and information costs.[135] Under the payment system provided by E-SDR, all trading nodes will have access to comprehensive and actual market data and information based on a free and fair, and self-regulatory system. This is because

---

[131] Müller et al. [137].

[132] Qian [138].

[133] Rahman [139].

[134] Vigna and Casey [140].

[135] Teramoto and Jurcys [141].

E-SDR is not only a settlement medium but also an open and transparent transaction system that can be participated by multiple parties. According to Donato Masciandaro, a professor at Bocconi University in Italy, digital currencies will have the function of information storage in addition to the medium of exchange and value storage functions of traditional currencies. The information store will then enable the currency to play more of a data exchange role, which will help lubricate resource allocation and improve market efficiency in a state of financial disintermediation.[136] Under E-SDR, all transaction records, asset information, and other data are recorded in a distributed form, transactions between nodes are updated and verified in a distributed accounting method, and users of all nodes can access and view this information at any time after obtaining authorization. This means that E-SDR can build a complete information connection channel for both sides of the transaction beforehand and complete the tracking and recording of the transaction in time afterward. Therefore, E-SDR can somewhat reduce the reliance on intermediaries in economic activities of various subjects and reduce transaction friction, coordination and communication costs of all parties, and large amounts of capital deposits caused by the presence of third-party intermediaries.[137] This will not only help break the protection of interests caused by centralized data management but also enable efficient information exchange and capital transactions, thus realizing "multiple scenarios with one account, multiple accounts with one platform, and gradual super account nation of platforms" on a larger scale, and promoting the development of economic paradigms toward digitalization, networking, and technology.

Fourth, E-SDR can build a new means and system of credit proof for the financial system. According to Narayana Kocherlakota, a professor at the University of Rochester, the update of credit information under the traditional financial system is often lagging and static. This leads to a credit system that does not fully play its role as a value support and greatly restricts the optimal allocation of resources.[138] E-SDR can automatically and completely record all transactions in the credit ecology through program algorithms, and store all books and copies in a distributed and encrypted manner on multiple terminal platforms, making it easier for all participants to check credit information and assess the credit status of the transaction objects. Therefore, based on establishing a complete and usable information transfer mechanism, E-SDR can realize the traceability of asset flow with blockchain depository and

[136] Masciandaro [142].

[137] The central payment clearing systems widely used worldwide at now are SWIFT and CLS (Continuous Linked Settlement). CLS adopts the "Payment to Payment" method for continuous synchronized settlement. Both systems are still mainly applicable to sovereign national currencies (virtual digital assets are excluded), and there are strict limitations on the users who can access them (mainly medium and large banks, financial institutions, and multinational corporations). At the same time, these two systems need to go through agents to establish relationships in cross-border settlement, so there is an unavoidable delay in time and payment speed (about 2–3 days for SWIFT and 1–2 days for CLS), as well as a certain amount of handling and service fees (about 7% of the transaction amount for SWIFT and a fixed monthly fee for CLS). In addition, both systems take a centralized approach to payments and, therefore, also run the risk of leaking trader and transaction information. Mägerle [143].

[138] Kocherlakota and Wallace [144].

provide a more credible environment for all kinds of transactions on a technical level. In addition, E-SDR can embed transaction rules such as laws and regulations and commercial contracts into its settlement system and perform intelligent execution of commercial rules through its consensus protocol and smart contracts, thus reducing the marginal cost of transactions and facilitating cross-border payment and settlement business transactions. In this regard, Khaled Salah, a professor at Khalifa University of Science and Technology in the UAE, and others believe that the involvement of smart contracts in digital currencies will not only provide an effective means of enforcing business rules but will also provide a viable solution for monetizing IoT data.[139]

Fifth, the digital form and systemic independence of E-SDR will expand its payment network. It will give domestic retail and small-scale users a way to clear small payments. This meets the growing need for flexible, fragmented payments and transactions by that user group. As a result, E-SDR is also highly valuable for highly fragmented transactions such as micro and small payments, sales terminal payments, online shopping, and domestic and international transfers. Gerald Dwyer, a professor at Clemson University, said that digital currencies could provide more convenient payment and settlement tools for peer-to-peer transactions, thus fully improving the liquidity of market participants.[140] At the same time, Wim Raymaekers, head of SWIFT's Global Payments Innovation Program, believes that the requirements of transactions for data transmission and processing efficiency will drive digital currencies to adopt a unified or similar technical architecture and value proposition. Therefore, if widely used as a digital system, E-SDR will also further improve the unity of financial infrastructure, which will enhance the effectiveness of financial infrastructure and reduce the cost of investment and maintenance efforts in systems and operations at all levels of platforms.[141] In addition, E-SDR has wide applicability, and various entities can access the system according to certain categories of affiliation. E-SDR also enables all market participants to maintain the stability and continuity of the system, thus improving the integrity and reliability of financial interests and transaction information. Claus Dierksmeier, a professor at the University of Tübingen, Germany, pointed out that the polycentric and multi-party participation of the digital currency system can break the traditional oligarchic centralized system and build an inclusive ecological environment for the sustainable development of financial inclusion.[142]

Sixth, the vertical and flat digital financial system built by E-SDR will also greatly improve the survivability of the market and, thus, more effectively combat money laundering, tax evasion, corruption, and other illegal acts. The digital currency system based on blockchain technology, as a secure encrypted, self-trusting, and traceable

---

139 Salah et al. [145].

140 Dwyer [146].

141 For research related to blockchain and digital currencies and financial infrastructure see Jun [147], Walch [148].

142 Dierksmeier and Seele [149].

form of money, is naturally applicable to the real-name mechanism.[143] E-SDR's digital financial account system can realize highly complete real names of each node through the design of relevant algorithms and mechanisms based on "clustering." At the same time, the digital financial system built by E-SDR enables all data of economic activities to be recorded and tracked so that regulators can analyze data such as the source and destination of funds, the reason for payment, the amount of payment, and the frequency of payment after obtaining authorization, and thus more fully grasp the basic situation of market operation and detect abnormal trading activities in a more timely manner. Gerald Dwyer, a professor at Clemson University in the United States, has fully affirmed the role of blockchain intervention in the currency layer to improve the ability of regulators to regulate the market.[144] Therefore, based on E-SDR, central banks, and other regulators will have better regulatory capabilities.

The Ripple system for payments and clearing based on the concept of multi-centralization, for example, is based on its own Ripple Consensus Protocol, which can automatically receive the general ledger transaction records from all nodes accessing the system and verify all types of transactions in the system, thus building an efficient settlement system and distributed bookkeeping system.[145] In terms of composition, Ripple is composed of Ripple coin (XRP), Gateway, and Ripple Payment Protocol (RTXP). To use it, users must open a Ripple account (20 XRPs need to be in the account). After obtaining the account's private key, the user can set up a gateway trust and make payments and transactions in the gateway (Fig. 4.9).[146]

Ripple's access threshold is relatively low compared to SWIFT, CLS, and other settlement systems (almost all financial institutions can access it), so it has a relatively wide range of participants. At the same time, Ripple's settlement function is distributed, enabling instantaneous, peer-to-peer payments and transfers without the involvement of intermediary financial institutions such as clearing centers and

---

[143] Wei et al. [150].

[144] Dwyer [151].

[145] The Ripple consensus algorithm contains two mechanisms: the "consensus mechanism" and the "verification mechanism." When the system generates a new instance of the ledger and generates a new transaction record, these two mechanisms will quickly verify the transaction, and the ledgers in the system are arranged in chronological order and linked together to form the total ledger of the Ripple system. It is worth noting that Ripple's consensus generation does not require the consent of all nodes but is accomplished by nodes in the system's trust list, i.e., as long as most of the nodes in this trust list express their agreement, the ledger can be deemed valid. See Wang Chaoyang and Bugao [152].

[146] Ripple (XRP) is a non-statutory digital currency based on the Ripple protocol and OpenCoin's original algorithm, with a total volume of 100 billion XRP. XRP serves as a bridge currency within the system, whereby users can exchange any type of currency or virtual currency for Ripple XRP and send it to any other user anywhere in the world. On the other hand, XRP serves as a security tool, each account needs at least 20 XRP to curb fraudulent transactions and massive transactions of malicious attacks. The gateway is the gate through which funds enter and exit Ripple's system (similar to currency access and exchange agencies), where transactions can be made peer-to-peer based on a trusted gatekeeper. RTXP is the Ripple payment protocol, which solves the problem of payments between different nodes (similar to the relationship between HyperText Transfer and Web Messaging, or between the Simple File Transfer Protocol and email). See Alexandre Mouradian, Isabelle Augeblum and Valois [153], Armknecht et al. [154].

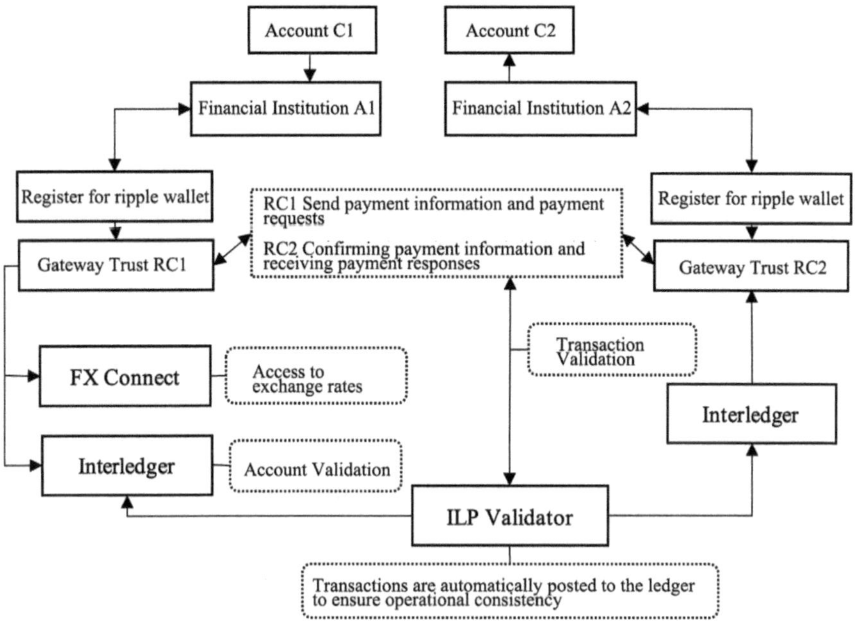

Fig. 4.9 Ripple System Communication Flow Diagram

banks.[147] In addition, although Ripple does not charge any fees for transfer services, each Ripple account must have at least 20 Ripple coins as a means of verification for normal transactions to ensure the system's safe operation.[148] Currently, Ripple can support atomic-level transactions in any currency worldwide, meaning that Ripple not only supports settlement between mainstream sovereign credit currencies such as the U.S. dollar, euro, Chinese yuan, and Japanese yen but also supports transactions in valuables, including digital currencies such as bitcoin and merchant credits. For sovereign credit currencies, Ripple supports foreign exchange transactions, cross-border transfers, and other types of business and can automatically convert exchange rates. Users can use any currency to pay others in any other currency, allowing for network-wide circulation of all currencies. Although the Ripple system still has defects such as doubtful distribution, low transparency of the platform, anonymization of funds, and poor traceability of transactions, its operation mechanism still provides new ideas and practical experience for constructing E-SDR.[149]

---

147 Keju [155].

148 Users consume one hundred thousandth of an XRP per transaction (as of March 16, 2019, the price of Ripple is 0.31960 against the US dollar), but for a malicious attacker who creates massive amounts of fake account and transaction information, the amount of XRP it needs to destroy grows geometrically due to the growth of transactions. Source: Investing, XRP/USD , https://cn.investing.com/crypto/ripple, accessed on March 16, 2019.

149 The market's skepticism about Ripple has been focused on XRP and its parent company, Ripple Labs. The XRP has adopted a "trust-based alternative" that consists of electing a small number of

Of course, the E-SDR system conceived in this paper with blockchain technology as its core has some features and functions beyond the guidance of existing monetary theories, and the E-SDR also still lacks measurable financial structures and indicator data. Therefore, there is still a need to conduct sufficient theoretical interpretation and empirical simulation on the rule design and feasibility of E-SDR. But what is more important to recognize is that the globalized connection platform formed in the Internet era has laid a sufficient foundation for establishing a highly efficient and low-cost global payment and clearing system. This also means that E-SDR has certain possibilities of realization. It is foreseeable that with the improvement of technology and system, the construction of the E-SDR super-sovereign digital currency system will be more feasible. And the realization of this system will vigorously promote the transformation and upgrading of global economic governance.

4.5 A New Framework for Global Economic Governance Under E-SDR

A distinctive feature of the current global economic development is that developing countries, represented by China, have maintained rapid economic growth, and these countries will further provide most of the support for the development of globalization. At the same time, the global economic governance system has failed to make corresponding adjustments to this new development in a timely manner, and the original systems, rules, and procedures are facing ineffective or inefficient problems.[150] However, with the new technological conditions and the joint efforts of all countries, global economic governance and the global order can still be transformed.

The root cause of the "center-edge" order is the unequal international economic division of labor and exchange system caused by the globalized capitalist production system. The "center-periphery" structure maintained by the current international monetary system is the most direct manifestation of this system. The dominance of this order has led to the formation of a global pattern of "dominance-dependence". The developed countries in the center dominate the institutional rules of global economic governance to safeguard their vested interests and check emerging economic forces' rise. To a certain extent, this has led to a huge imbalance in development between countries and further intensified international competition and conflict. In this regard, Yale University Professor Thomas Pogge said that the existing global order represents more of the interests of developed countries.[151] At the same time, the development of globalization requires the construction of a more dynamic and

nodes to form a "Unique Node List." However, the selection special nodes is at the discretion of Ripple Labs. Meanwhile, XRP exists independently of Ripple Labs. However, Ripple Labs holds many XRPs, and this company has not publicized the distribution of its Ripple coin stock or its sales. In addition, XRP is not used by its financial partners. See Caplinger [156].

[150] Deudney and Ikenberry [157].

[151] Pogge [158].

inclusive global economic governance system on the one hand and the full mobilization of the global flow of innovative resources on the other.[152] The "center-periphery" order cannot meet the needs of the highly interdependent global economic landscape. Moreover, because of the contradiction between global economic governance and internal governance of each country in terms of universality vs. specificity, homogeneity vs. heterogeneity, and weak politics vs. strong politics, a structure of the interest-based relationship is still maintained between developed and developing countries. The dependency and exclusion between the "center" and the "periphery" remains in place. The logic of dependence and exclusion between the "center" and the "periphery" will not be weakened in the near future.[153] For this reason, building a cooperative and inclusive global economic order with justice based on common interests is an important way out of the current global economic governance dilemma.

The E-SDR based on blockchain technology can break the "center-periphey" order and provide a new perspective for transforming global economic governance. In fact, since the essence of money is a consensus-based credit, the monetary competition is a competition for consensus credit. The establishment of the current international monetary system is the result of national sovereign credit competition, i.e., the dollar-centeredness of the current international monetary system is that the credit of the United States has partially or fully replaced the credit of other countries.[154] The dollar-centered international monetary system not only exacerbates the problem of structural imbalances in global economic governance but also leads to recurrent instability in global economic development.[155] As a theoretical exploration and financial practice of a super sovereign monetary system, the E-SDR is built based on digital credit realized on blockchain and other technologies and their supporting mechanisms. At the same time, the E-SDR is issued based on the data of economic growth and social goods production, which is calculated by a series of intelligent algorithms; that is, the E-SDR can make timely and effective adjustments according to the fluctuation of social production conditions or economic production efficiency. This digital credit and price adjustment mechanism can break the limitations of traditional metal commodities and specific sovereign currencies in terms of monetary anchoring to a certain extent. This means that the formation mechanism and issuance of E-SDR depend on the total amount and allocation of resources in the real economic system and are more closely integrated with the development of the global economy.

In addition, under the monetary system anchored to national credit, the money supply is in a state of long-term expansion, and the global issuance discipline does not fully bind the global circulation of sovereign credit currencies. This persistent expansion not only breaks the internal resident anchor system where the value of money is

---

[152] Yuyan [159].

[153] Zhicheng [160].

[154] McKinnon [161].

[155] Under the current international monetary system, most of the international reserve currency supply comes from the United States' domestic credit expansion, and the United States exports the "rules of the game" to non-reserve currency countries through trade deficits. See Daoping and Xiaoyun [162].

anchored by government credit but also exacerbates the mismatch between monetary aggregates and government credit aggregates, leading to significant imbalances in exchange rates and inflation expectations.[156] Thus, as David Harvey, a professor at the City University of New York, argues, the expansion of unconstrained government credit has exacerbated the chaos in the international monetary system.[157] On the other hand, the E-SDR system is constructed on the basis of multi-party participation, mutual monitoring, and joint maintenance. This is because the E-SDR is constructed through a distributed network as a multi-way trust relationship, i.e., the formation of a global credit network is based on the joint participation of countries.[158] On the premise that this trust network is realized, E-SDR can better embed the homogeneous basis and heterogeneous needs of multilateral subjects in monetary adjustment.[159] From this perspective, E-SDR can give the currency a broader social character by reducing the privacy of individual national centers, thus avoiding inflationary and benefit redistribution tendencies caused by financial hegemony or administrative intervention.[160]

Of course, the result of E-SDR development is not a simple decentering but more likely a breaking of the single centrality in a polycentric or weakly centralized way. Indeed, an important manifestation of the change in the intrinsic dimension of money on social relations lies in the transformation of individual single-point connections into multi-point connections of universal interactions.[161] When the social state of the interconnectedness of everything is realized, all social information will form an all-around interactive relationship, and the development and utilization efficiency of human, material, and financial resources will be greatly enhanced.[162] Under this social state, the total factor productivity and the utilization rate of social resources can be optimized to the greatest extent. At the same time, this state of society can also enhance the correlation of rights, responsibilities, and interests among countries. This will help transform the governance model of national relations into a consultative governance structure and promote the emergence of a sense of human destiny community.[163] But this holistic and global participation is not intended to completely overturn the previous centralized economic governance pattern but to

---

[156] Tao et al. [163].

[157] Harvey [164].

[158] Yide and Mingqi [165].

[159] Lei [166].

[160] Jinhan and Jian [167].

[161] Xiong [168].

[162] In the Economic Manuscripts of 1857–1858, Marx states, "The social connection of the individual manifests itself in exchange-value, for it is only by means of exchange-value that, for each individual, his own activity or product becomes his activity or product... Each person's ability to control others or community wealth comes from owning exchange-value or money. He carries his social power and connection with society in his coat pocket." Indeed, Marx saw the creation of money as a manifestation of the development of economic and productive relations and an essential driver of socially active interaction. See *The Complete Works of Marx and Engels: Volume IV*, Beijing: People's Publishing House, 2007, pp. 106–107 (in Chinese).

[163] Changhe [169].

promote the transformation of this pattern into an open, differentiated, and service-oriented polycentric economic governance pattern.[164] Such a transformation will effectively guarantee equal rights among countries and achieve a more reasonable and just order.[165] Therefore, realizing the polycentric transformation of the global economic governance pattern is also the key to building a global just economic order.

Clearly, E-SDR can facilitate this transformation by building a system of socially valuable connections. As Primavera Filippi, a researcher at the French National Center for Scientific Research, says, "What blockchain is trying to achieve as an institutional technology is a global language of cooperation."[166] On top of accelerating the transformation of the international monetary system, the E-SDR will also expand the space for countries to participate in global economic governance. Based on the digital mode of operation, E-SDR can form a smoother monetary transmission mechanism, improve the liquidity of the operation of the overall economic system of society, and thus reduce the restrictions of the original sovereign credit currency system on countries' participation in global trade and regional value chain division of labor.[167] In this regard, American economist Richard Levin and his research team pointed out that digital currencies can achieve the maximum diversity of currency users thanks to the blockchain's polycentric mechanism.[168] At the same time, E-SDR can improve the economic operation mechanism and market self-regulation environment and achieve a more fair and efficient social benefits distribution, leading to a more adequate and balanced distribution of global economic development and related resources. According to Estonian scholar Alex Pazaitis, blockchain and digital currency can promote the transmission and sharing of information flow on a global scale by creating a value coordination mechanism and public-oriented ecosystem, thus strengthening social cooperation on a global scale and realizing the efficient use of overall resources.[169] Such diversified participation and equitable distribution will contribute to forming more equal and inclusive universal rules for global economic governance, thus providing a beneficial environment for various social actors to participate in economic globalization. In addition, E-SDR will accelerate the global flow of innovative factor resources and promote the expansion of trade, commodity, and information flows with low transaction cost incentives, thus realizing a new

[164] In this regard, Xi Jinping believes that "in the era of economic globalization, the development of all countries is intertwined. We rise or fall together. No country can do well independently, and coordination and cooperation is an inevitable choice. We must realize linked development in the resonance of the world economy." See "Building an Innovative, Dynamic, Linked and Inclusive World Economy—Opening Speech at the Hangzhou Summit of G20 Leaders," in *People's Daily*, September 05, 2016 (in Chinese).

[165] Equal rights refer to the fact that all countries, regardless of their size, strength, or weakness, should be treated equally and fairly. A just order refers to mutual benefit in the economic field, fair allocation of resources, recognition of cultural equality, and a necessary balance in politics and even military affairs. See Charles Beitz, translated by Zhanxiu [170].

[166] Filippi [171].

[167] Tapscott and Tapscott [172].

[168] Levin et al. [173].

[169] Pazaitis et al. [174].

globalization mechanism driven by economies of scale and digital economy. Mark Dodgson, a professor at the University of Queensland in Australia, and David Gann, a professor at Imperial College London in the U.K., among others, have proposed that digital currencies themselves will not only be able to measure and store value digitally but also enable faster, easier, and more widespread transactions, thereby breaking down temporal and spatial constraints on the use of funds and reducing friction in the financial system.[170]

Although the E-SDR super sovereign currency system proposed in this paper is not complete, the concept is still of great importance to China from the following three perspectives: First, from the perspective of RMB internationalization, although the transfer and decentralization of monetary power is a fundamental feature for the transformation of the global political and economic system, and RMB has already gained a certain international status in the current international monetary system. The RMB has already gained a certain international status in the current international monetary system. However, as a new force outside the system, this transition process and RMB internationalization cannot be a straightforward path without obstacles.[171] In fact, the international monetary system itself is highly political and strategic as a model and path to shape the distribution of economic power and benefits. This nature determines the existence of a certain degree of confrontation in monetary relations among countries, leading to a more intense game between countries (especially large countries) around their national interests.[172]

At the same time, some countries in Europe, East Asia, and OPEC still maintain a close alliance with the United States. This one-way dependence on national security also makes it difficult for China to unite these countries in cooperation and coordination to promote reform of the current international monetary system.[173] In this regard, Michele Fratianni, a professor at Indiana University, points out that geopolitics is an important influence on the international community's difficulty in effectively adjusting the imbalances in the international monetary system. At the same time, the convergence of real interests has made it possible for the currencies of Western developed countries to maintain an absolute advantage in international monetary competition.[174] Therefore, at present and for a long time, the dollar and the euro in the Western capital system will remain dominant, and it is still difficult for China to adopt effective external checks and balances to weaken or suppress its hegemonic position.[175]

In addition, there are still major disagreements among countries about the rebalancing of international relations and the alternation of power, and more and more

---

[170] Dodgson [175].

[171] Hongyu and Xiaosan [176].

[172] Confrontation is an intrinsic attribute of international monetary relations, and as long as the international system is still composed of sovereign states, inter-state monetary confrontation centered around national interests will exist for a long time. See Ping and Qingyou [177].

[173] Wei [178].

[174] Fratianni [179].

[175] Helleiner and Kirshner [180].

countries are interacting in an exclusive and confrontational manner. This suggests that the logic based on power politics that existed under the original centralized global economic governance system continues to play a profound role in the governance process. As Richard Haass, president of the Council on Foreign Relations, has argued, despite the declining hegemonic dominance of the United States globally, a cooperative language that can address global challenges has yet to emerge.[176] The imbalance in development levels among countries, differences in national interests, and geopolitical crises have created significant constraints on monetary cooperation.[177] Therefore, although there is a positive situation of interdependence and co-promotion between the RMB and other international currencies, the internationalization of the RMB will inevitably involve currency competition and currency substitution and affect the established pattern of the existing international monetary system. Suppose we continue to follow the traditional path of currency internationalization to push forward the globalization of the RMB and thereby promote the reform of the international monetary system. In that case, this will likely increase the resistance of other countries to China, which will, in turn, prevent China from cooperating globally and promoting the reform of global economic governance.

Second, from the perspective of the current U.S.-China relationship, the United States was originally the global development leader, while China was the global development follower, and the two were at unequal stages of development. However, with the deepening of globalization and structural changes in the global governance system, China's economic aggregates and comprehensive strength have formed a more solid foundation. The United States, on the other hand, has entered a period of contraction of its hegemony, which has led it to adopt a more self-interested and exclusionary policy orientation.[178] As Bruce Russet, a professor at Yale University, has argued, while the power base of U.S. hegemony has begun to decline, hegemony as a result of control has not.[179] Thus, the tendency of power shift and balance between China and the United States in various regions has also become a reality. But as the leader of the current system, the U.S. will inevitably find it difficult to accept the reality of China's leap in comprehensive power in the short term and will inevitably become more concerned and uneasy about the changing global order.[180] At the same time, most of the reasons why the U.S. starts wars or trade wars abroad are related to the shaking of the U.S. dollar's hegemony, and the U.S. has implemented a high-handed strategy of prevention and containment against the rise of other currencies, such as the yen and the euro.[181] In fact, in order to maintain its hegemonic position,

[176] Haass [181].

[177] Xiao [182].

[178] Guoyou [183].

[179] Russet [184].

[180] Long-term, routine, and fluctuating conflicts and games are very likely to become the norm in Sino-American relations in the future. However, the competition between China and the United States should be benign, positive, and transcendent rather than mutual replacement and zero-sum game competition. China and the U.S. should use competition to promote cooperation and use cooperation to create a win-win situation. See Guoyou [185].

[181] Xiao and Yongqi [186].

the U.S. tends to adopt a more aggressive policy in foreign relations, and this is particularly evident in foreign trade and economic relations.[182] Thus, with the shift in the U.S. self-interest and exclusionary policy orientation, the dollar hegemony also gradually began to change to coercive hegemony.[183] Additionally, imbalances in domestic U.S. politics and interest groups expanded economic and trade issues between China and the U.S. This further escalated into national strategic competition, with trade frictions marking the beginning. The pattern of strategic competition between China and the U.S. began to take shape. Against this backdrop, if the RMB internationalization continues along the original path, it will most likely expand the competition between the RMB and the USD and even lead to a complete zero-sum game between the two sides. However, the core of economic governance lies in the currency, and once the zero-sum competition between the RMB and the USD emerges, it will most likely intensify the conflict between the U.S. and China in economic and trade matters.[184]

Third, from the dimension of contemporary China's international positioning, China, as a rational and responsible member of the international community, should become not only a participant and cooperator of the existing international system and order but also a shaper of a more just and reasonable international order.[185] As Xi Jinping said, "To promote an open, inclusive, inclusive, balanced and win–win economic globalization, we should not only make a bigger cake but also share it well."[186] Unlike the West, which opposes technology sharing and mechanism co-construction, China places more emphasis on sharing and common progress with other countries in technology development and cooperation and seeks to assert and defend the interests and aspirations of developing countries.[187] In terms of global economic governance and its reform, China has also been adhering to the principle of co-business, co-build, and co-share, promoting the development of a more just and reasonable global economic governance system, thus creating more favorable conditions for building a new type of international relations with win–win cooperation as the core, promoting common development and building a community of human destiny.

To this end, I believe that in the process of RMB internationalization and monetary regional cooperation, China also needs to develop a monetary cooperation route that meets the common interests of most countries around the world and to promote the reform of global economic governance in a controlled economic order as much as possible, so as to avoid political confrontation due to economic conflicts. And E-SDR is an intermediate route of monetary cooperation. On the one hand, the E-SDR itself is constructed by the stakeholders of the international community through joint efforts, so the E-SDR has the ability to integrate the interests of most sovereign

182 Kindleberger [187].

183 Snidal [188].

184 Mingli [189].

185 For reflections on China's international positioning, see Tuo [190].

186 Jinping [191].

187 Qiqi [192].

countries into the new monetary framework and thus helps avoid the intensification of international competition due to currency competition. On the other hand, the E-SDR can build an international platform and trust mechanism for the interaction of resources and information for global economic governance, thus strongly promoting global economic interconnection, sharing, and governance and effectively coordinating the values, interests, and needs of all parties.[188] E-SDR can create a flexible space and buffer for China to participate in the reshaping of the international monetary system and promote the change of global economic governance. Meanwhile, China can also use E-SDR to strengthen cooperation with other countries. What's more, it can help China build a participatory leadership in global economic development and accelerate the transformation and upgrading of global economic governance.

At present, digital economy forms such as mobile payment, e-commerce, and platform economy have to some extent, laid a certain foundation for the development of super-sovereign digital currency, and some countries have started to try to build digital currency as a response to the development of the monetary system.[189] However, the acquisition of E-SDR-related resource endowments needs to be built on the basis of a highly robust digital ecology and supported by well-established capital and technology-intensive industries. As a result, E-SDR has higher requirements for financial infrastructure and its supporting macro policies. In terms of the hardware environment, E-SDR needs to configure and upgrade corresponding facilities and equipment and build and expand various application scenarios to ensure the stability and security of use. Regarding the software environment, E-SDR also needs to achieve coordination in various aspects such as technology, practice, and regulation and complete the work of system construction, policy support, and promotion and publicity to ensure coordination and widespread use. More importantly, since blockchain technology has a certain system integration effect and economy of scale effect, the prior advantage will be more obvious in E-SDR construction. This capability advantage will likely lead developed countries to continue occupying the high ground of blockchain technology. In contrast, developing countries lose their comparative advantage.

Since developed countries have world-class financial and information industries, they have an absolute advantage in the innovation and application of digital currency and other areas of blockchain. This has been reflected in the blockchain 1.0 tier applications, i.e., the digital currency activities in developed countries and regions are significantly better than those in other regions.[190] Large emerging countries that

---

[188] Ruguo [193].

[189] For example, the Monetary Authority of Singapore has undertaken the Ubin project to process cross-border transactions in real-time with central banks around the world through blockchain technology; the European Central Bank and the Central Bank of Japan have jointly undertaken the Stella project to build a complete Delivery versus Payment (DvP) system in a distributed ledger environment; and the Central Bank of Canada is leading the Jasper project to build an "Integrated Securities and Payments Platform" through the distributed ledger technology of blockchain.

[190] According to the Global Digital Cryptocurrency Benchmarking Study published by the Cambridge Centre for Alternative Finance (CCAF) in the UK, most of the 51 digital cryptocurrency exchanges in 27 countries it surveyed are located in Europe and the Asia-Pacific region. Meanwhile,

have established a certain industrial system can grasp the development trend of blockchain technology in a more timely manner and gradually participate in the construction of E-SDR because of their relatively complete industrial structure and better digital information foundation. For small and medium-sized developing countries that rely on resource exports and cheap labor, they are likely to face the problem of "participation deficit" in the financial, technical, and institutional aspects of E-SDR construction due to the imbalance of their industrial composition and the lack of real industries to support the development of the digital economy.[191] Suppose developing countries (especially small and medium-sized developing countries at the early stage of development) are not able to grasp the development of global blockchain technology in a timely and effective manner. In that case, they will most likely lose the ability and power to participate in the construction of E-SDR, and will continue to be shackled by the international pattern of "center-periphery."[192]

Taking the recent emergence of Stable Coin in the international market as an example, the New York State Department of Financial Services (NYDFS) has approved two stable tokens based on Ether ERC20 on September 10, 2018—Gemini Dollar issued by Gemini Trust and Paxos Standard by Paxos Trust.[193] Although NYDFS mainly provides credit guarantees and is not directly involved in issuing and operating specific dollar-based stable tokens. However, as the official government of the United States, the endorsement of NYDFS will bring a strong credit enhancement effect to such dollar-anchored stable tokens, enabling them to gain a larger market share.[194] More importantly, it also means that NYDFS and the U.S. government are starting to establish regulatory rules for stable tokens to gain control over global stable token platforms. Of course, whether stable tokens will be able to penetrate the world and eventually become a payment currency is subject to further discussion. But the U.S. dollar itself is already the unit of denomination for most of the world's digital currencies. It is also one of the few fiat currencies that can be

85% of Asia-Pacific digital currency exchanges are unlicensed, while 78% of all North American exchanges have a formal government license or authorization. In addition, about 81% of the world's wallet services for digital cryptocurrencies come from North America and Europe, but these two regions account for only 61% of the total global user base. See Hileman and Rauchs [194].

[191] The "participation deficit" refers to the fact that developing countries often find it difficult to participate effectively in global governance because of the inability of countries with weaker national governance capacity to participate in global governance themselves and because the existing global governance order does not give emerging countries a corresponding voice in rule-making. See Xuelian and Lu [195].

[192] Qiqi [196].

[193] Stabilization tokens are digital currencies that aim to maintain parity with fiat currencies by "anchoring" to the fiat currency system or digital assets. The most widely used stabilization token is USDT, issued by Tether, anchored to the US dollar at a 1:1 ratio. However, the core requirement of this type of anchored fiat-collateralized stable tokens is a 100% fiat deposit as the issuance reserve and a guarantee that the stable tokens and the anchored fiat currencies can be cashed out at par at any time. Therefore, its substance is more of a depository receipt for non-financial institutions, which differs from the E-SDR proposed in this paper. See Wang and Li [197].

[194] Castor [198].

exchanged with digital currencies.[195] It is not difficult to see that the United States has already started to make a pre-emptive layout in the digital currency field based on its first-mover advantage in key areas such as financial mechanisms, technology, and talent.

China has established a relatively complete industrial system, and the influence of RMB in the international monetary system is gradually increasing. For example, to further integrate RMB cross-border clearing channels and improve the efficiency of RMB cross-border payment settlement, the People's Bank of China has built the Cross-border Interbank Payment System (CIPS), which has also basically achieved full coverage of the global market.[196] Therefore, China has the opportunity and ability to actively participate and promote the construction of future digital currencies. Of course, although China has constructed the initial development strategy of blockchain and the positive value of blockchain technology is gradually emerging, China still lacks the underlying strategy of specific blockchain development, and the industrial development and industry regulation policy system are yet to be improved. Specifically, the development of blockchain technology in China mainly has the following problems: (1) Blockchain technology lacks interfaces with the existing financial infrastructure. It cannot effectively communicate and interact with data from the non-blockchain world, and it also does not sufficiently cross and integrate with related technologies and fails to export powerful computing power through collaborative computing. Therefore, the network-wide computing power of blockchain technology has not yet formed a systematic value and remains at the stage of internal competitive computing to maintain its operation. The phenomenon of "value island" caused by "information island" is still serious.[197] (2) Blockchain technology itself is still in the early stage of development, and it is still difficult to break through the application scenario and realize the product landing. Moreover, the digital cryptocurrency represented by Bitcoin has consumed too many social resources in the early stage, and other blockchain-related infrastructure has not been paid attention to effectively, which hinders the landing of blockchain technology at the application level and the formation of a complete ecosystem. (3) Blockchain technology has not yet been tested in a large-scale transaction environment, and its resilience and supervisability are still in doubt. Moreover, the core foundation of blockchain technology is still in the hands of developed countries such as Europe and the United States, and China has not yet been able to form systematic theories and applications at the basic technology level.

For this reason, China should seize the opportunity of technological development and grasp the laws of economic trends, make adequate preparation for conceptual precursors and technical support, and establish a digital currency framework

---

[195] Galicki [199].

[196] CIPS (Phase I) was successfully put into operation on October 8, 2015, while CIPS (Phase II) was officially launched on May 2, 2018. As a milestone event in the process of RMB internationalization, CIPS not only reduces China's dependence on the SWIFT system, but also improves the internationalization and security of China's financial system. See Zhong [200].

[197] Kuo Chuen [201].

and industrial layout that meet the requirements of modernizing national governance capacity based on an in-depth understanding of the main characteristics and external constraints of blockchain technology. We must strengthen the research of blockchain-related basic theories, increase the investment in blockchain technology infrastructure, solve the problems of capital and technical threshold for its development, and provide strong guarantees for industrial blockchain projects to serve the real economy so as to provide sufficient technical capacity and industrial strength for the construction of E-SDR. At the same time, we must also strengthen the supervision of the blockchain industry, accelerate the development of key technical standards of blockchain and improve the industry standards of blockchain, prevent the potential risks of blockchain and financial chaos caused by technological innovation, and avoid speculators from using the concept of blockchain for financial fraud and other activities, thus promoting the standardized and healthy development of blockchain industry and maintaining the stability and security of national finance. In addition, the lack of capacity of many developing countries to build an autonomous international monetary network and the reluctance of developed Western countries, which have traditionally dominated global economic governance, to provide more public goods may lead to low motivation for countries to participate in the E-SDR. Therefore, China also needs to promote the unification of digital standards and the integration of cooperation mechanisms, actively promote the open source and sharing of the underlying blockchain technology, and form a new open language in the development of blockchain technology and the construction of the E-SDR, so as to buy time and space for developing countries' development in the digital currency field to maintain its leading edge in the digital asset field and gain participatory dominance.

4.6 Conclusion

The current acceleration of globalization is driving a shift from an international system based on the great power landscape to a global system based on the interdependence and interaction between developed and developing economies.[198] The center of global economic governance has thus been transformed from a system of sovereign states to a new global society with shared vulnerability and responsibility.[199] In this transformation, global economic governance is challenged by systemic and structural imbalances. This institutional imbalance refers to the fact that the system of global economic governance and the economic order that results from its implementation are the result of a deliberate plan pursued by a small number of individuals. Therefore the system is necessarily predisposed to certain explicit intentions. With human intervention and manipulation, these institutional designs inevitably become unbalanced. The current hegemonic dominance in global economic governance is a typical example of institutional imbalances. Structural imbalance occurs when the

[198] Yiran [202].

[199] Cotton [203].

structure of social production becomes uneven because the economic structure cannot adjust quickly enough in response to market changes. This imbalance is reflected in the institutional inertia and path dependence in global economic governance. The key to resolving these two imbalances lies in the full flow of market information, the participation of multiple parties in consensus decision-making, and limited precision regulation.

As a framework underlying technology, blockchain technology will play an important role. Blockchain can realize the construction of polycentric architecture and distributed transactions, maximize the flow and sharing of existing data and information, and thus expand the boundaries of cooperation and improve the efficiency of risk management. At the same time, blockchain can create good technical conditions for constructing self-trust ecology and create a new type of free and open system collaboration mechanism. More importantly, blockchain can form technical resonance with artificial intelligence, i.e., blockchain can provide more value-creation opportunities for the application of artificial intelligence based on data sharing and ensuring data security. Therefore, embedding blockchain technology can help build a polycentric public information platform with borderless value circulation and release new economic values by promoting the formation of a new language of social cooperation. This will help solve the institutional and structural imbalance in global economic governance and thus provide a new perspective and realization path for transforming and upgrading global economic governance.

References

1. Cooper R (1975) Prolegomena to the choice of an international monetary system. Int Organ 29(1):65
2. Gilpin R (2001) Global political economy: understanding the international economic order. Princeton University Press, Princeton, 248–250
3. Wei L (2012) The political foundation of balancing against dollar: the rising economy responds to the U.S. currency hegemony. World Econ Politics 5:99. (in Chinese)
4. Steil B (2013) The battle of Bretton woods. Princeton University Press, Princeton, pp 334–335
5. Eichengreen B, Flandreau M (2009) The rise and fall of the dollar, or when did the dollar replace sterling as the leading reserve currency? Europ Rev Econ History 13(3):398–401
6. Gavin F (2011) translated by Yan Rong, Gold, Dollars, and Power: the politics of international monetary relations: 1958–1971. Social Science Academic Press, Beijing, pp 146–154. (in Chinese)
7. Mandel E (1969) The crisis of the international monetary system. Int Socialist Rev 30(2):147–151
8. Hogan M (1989) The Marshall plan: America, Britain and the Reconstruction of Western Europe, 1947–1952. Cambridge University Press, Cambridge, pp 12–24
9. Eichengreen B (2007) Global Imbalances and the Lessons of Bretton Woods. Econ History Rev 60(3):645–646
10. Spiro D (1999) The hidden hand of American hegemony: petrodollar recycling and international markets. University PressCornell University Press, Ithaca: Cornell, pp 344–345
11. Farrell D, Lund S (2008) The new role of oil wealth in the world economy. The McKinsey on Finance, New York, pp 2–4

12. Hudson M (2008) translated by Ji Fei and Lin Xiaofang, Super imperialism: the economic strategy of American empire. Central Compilation and Translation Press, Beijing, p 33. (in Chinese).
13. Guangbin Z, Shaozong Z, Yuanchang W (2017) Currency anchor options: historical review and current predicament. Int Econ Rev 6:77. (in Chinese)
14. Cooper R, Dornbusch R, Hall R (1982) The gold standard: historical facts and future prospects. Brookings Papers Econ Activity 1982(1):1–56
15. Xiangsui W (2017) A theory of currency-politics: the evolution of monetary Politics. CITIC Press Group, Beijing, pp 1–2. (in Chinese)
16. Keohane R, Nye J (2002) translated by Men Honghua, Power and Interdependence, Peking University Press, Beijing, pp 147–159. (in Chinese)
17. Hudson M (2003) Super imperialism: the origin and fundamentals of U.S. world dominance. J Int Studies 32(2):350–352
18. Mckinnon R, Schnabl G (2003) The east Asian dollar standard, fear of floating, and original sin. Comparat Econ Soc Syst 8(3):331–360
19. Xiangsui W (2011) Currency-politics: the changes and future of the world situations. World Econ Polit 4:15. (in Chinese)
20. Prasad E (2014) The dollar trap: how the U.S. dollar tightened its grip on global finance. Princeton Uiniversity Press, Princeton, pp 63–67
21. Yulu C, Fang W, Ming Y (2005) Currency Internationalization as a national competitive strategy: US Dollar's empirical evidence——and a study on the issue of Chinese Yuan. Econ Res J 2:37. (in Chinese)
22. Maurice O (1994) The logic of currency crises. In: Eichengreen B, Frieden F, Hagen J (eds), Monetary and fiscal policy in an integrated Europe, pp 85–86
23. Dailami M, Masson P, Padou J (2008) Global monetary conditions versus country-specific factors in the determination of emerging market debt spreads. J Int Money Financ 27(8):1333–1336
24. Krugman P (2007) Will there be a dollar crisis? Economic Policy 22(51):454–455
25. Dooley M, Folkerts-Landau D, Garber P (2010) Bretton woods II still defines the international monetary system. Pac Econ Rev 14(3):309–310
26. Umit O (2016) Stationarity of real exchange rates in the 'fragile five': analysis with structural breaks. Int J Econ Financ 8(4):254–255
27. Scott S, Zachariadis M (2014) The society for worldwide interbank financial telecommunications (SWIFT): cooperative governance for network innovation, standards, and community. Routledge Press, London, pp 187–188
28. Scott S, Zachariadis M (2012) Origins and development of SWIFT, 1973–2009. Bus History 54(3):471–473
29. Peyrou S (2015) Anti-terrorism struggle versus the protection of personal data: a European point of view on some recent transatlantic misunderstandings. J Policing Intell Counter Terrorism 10(1):49–51
30. Yuyan Z, Jingchun Z (2008) The nature of currency and the future choice of Chinese yuan—an essay on Asian monetary cooperation. J Contemp Asia-Pacific Studies No. 2:22 (in Chinese)
31. Kirshner J (2013) translated by Li Wei, Currency and coercion: the political economy of international monetary power. Shanghai People's Publishing House, Shanghai, p 302. (in Chinese)
32. Weixing X, Guannan W (2012) Sensitivity and vulnerability in the economic interdependence between china and the United States: an analysis based on the perspective of the 'financial terrorism equilibrium'. J Contemp Asia-Pacific Studies 6:92–93. (in Chinese)
33. Calvo G, Reinhart C (2002) Fear of floating. Quart J Econ 117(2):399–402
34. Prebisch R (1950) The economic development of Latin America and its principal problems. United Nations Department of Economic Affairs, pp 49–55
35. Prebisch R (1959) Commercial policy in the underdeveloped countries. Amer Econ Rev 49(2):251–273

36. Amin S (1974) Accumulation and development: a theoretical model. Rev African Polit Econ 1(1):9–26
37. Amin S (1997) Global restructuring and peripheral states: the carrot and the stick in Mauritania. Int J African Historical Studies 30(4):375
38. Preston P (2010) The other side of the coin: reading the politics of the 2008 financial Tsunami. British J Polit Int Relations 11(3):504–517
39. Marchi V, Giuliani E, Rabellotti E (2017) Do global value chains offer developing countries learning and innovation opportunities? Europ J Dev Res 30(3):405–407
40. Costigan T, Cottle D, Keys a (2017) The US dollar as the global reserve currency: implications for US hegemony. World Rev Polit Econ 8(1):117–119
41. Zhengyi W (2018) A theory of development in the margins: the world system and the Development of Southeast Asia. Shanghai People's Publishing House, Shanghai, p 10. (in Chinese)
42. Zhicheng W (2016) The influence of global governance upon state governance. Soc Sci China 6(6):25. (in Chinese)
43. Keping Y (2012) Trends in global governance and china's strategic choice. Foreign Theoret Trends 10:8. (in Chinese)
44. Ferson W, Harvey C (1991) The valuation of economic risk premium. J Polit Econ 99(2):386–389
45. Polanyi K (2013) translated by Huang Shumin, The great transformation: the political and economic origins of our time. Social Science Academic Press, Beijing, pp 210–212. (in Chinese)
46. Strange S (1985) Protectionism and world politics. Int Orga 39(2):254–256
47. Tuo C (2016) Global governance and state governance: two strategic considerations in contemporary China. Soc Sci China 6(8):8. (in Chinese)
48. Xiujun X (2012) Emerging economies and the structural transformation of global economic governance. World Econ Polit 10:76. (in Chinese)
49. Fan H, Weijiang F, Jin X (2013) On the challenges of global governance mechanism and China's countermeasures. World Econ Polit 4:20. (in Chinese)
50. Burda M, Dluhosch B (2010) Cost competition, fragmentation, and globalization. Rev Int Econ 10(3):437–439
51. Yu Z (2011) Global economic governance and China's participation strategy. World Econ Studies 11:29. (in Chinese)
52. Zhimin C (2016) State governance, global governance and the construction of world order. Soc Sci China No. 6:19 (in Chinese)
53. Zhengyi W (2010) General theory of international political economy. Peking University Press, Beijing, p 495. (in Chinese)
54. Yaqing Q (2013) Global governance failure and ideational reconstruction for a sustainable world order. World Econ Polit 4:6 (in Chinese)
55. Jing L (2014) The current institutional dilemma of global governance and its reform. Foreign Affairs Rev 31:112. (in Chinese)
56. Kissinger H (2011) Changes in the world economic governance system in the era of globalization. Rev Econ Res 49:5 (in Chinese)
57. Nakamoto S (2008) Bitcoin: a ppeer-to-peer electronic cash system. Bitcoin Organization, Available at: https://bitcoin.org/bitcoin.pdf
58. Antonopoulos (2014) Mastering Bitcoin: unlocking digital crypto-currencies. New York: O'Reilly Media, pp 195–211; Swan M (2015) Blockchain thinking: the brain as a decentralized autonomous corporation. IEEE Technol Soc Maga 34(4):41–52
59. Underwood S (2016) Blockchain beyond Bitcoin. Commun ACM 59(11):15–17
60. Aste T, Tasca P, Matteo T (2017) Blockchain technologies: the foreseeable impact on society and industry. Computer 50(9):18–28
61. Pierro M (2017) What is the blockchain? Comput Sci Eng 19(5):92–95
62. Iansiti M, Lakhani K (2017) The truth about Blockchain. Harv Bus Rev 95(1):9–10

63. Yong Y, Feiyue W (2016) Blockchain: the state of the art and future trends. Acta Automatica Sinica 4(4):483–484 (in Chinese)
64. Narayanan A, Bonneau J, Felten E, Miller A, Goldfeder S (2016) Bitcoin and cryptocurrency technologies: a comprehensive introduction. Princeton University Press, Princeton, pp 37–40
65. Restuccia F (2018) Blockchain for the internet of things: present and future. IEEE Internet Things J 1(1):6–7
66. Notheisen B (2017) Jacob Benjamin Cholewa and Arun Prasad Shanmugam, "trading real-world assets on Blockchain." Bus Inf Syst Eng 59(6):426–428
67. Christidis K, Devetsikiotis M (2016) Blockchains and smart contracts for the internet of things. IEEE Access 4(11):2299–2301
68. Dinh TTA, Wang J, Chen G, Liu R (2017) Blockchain: a framework for analyzing private Blockchains. Paper presented to the conference on "2017 ACM International conference on management of data," Chicago, May 14–19, 2017, pp 1087–1088
69. Christidis K, Devetsikiotis M (2016) Blockchains and smart contracts for the internet of things. IEEE Access 4(1):2293–2294
70. Dilleyet J (2017) Strong federations: an interoperable blockchain solution to centralized third party risks. CoRR abs, http://web.archive.org/20170922194157/https:/lightning.network/lightning-network-paper.pdf
71. Back A, Corallo M, Dashjr L, Friedenbach M, Maxwell G, Miller A, Poelstra A, Timón J, Wuille P (2014) Enabling Blockchain innovations with pegged sidechains. Blockstream, October 22, 2014, https://www.blockstream.com/sidechains.pdf.
72. Swan M (2015) Blockchain: blueprint for a new economy. O'Reilly Media, New York, pp 10–12
73. Swan M (2015) Blockchain: blueprint for a new economy. O'Reilly Media, New York, pp 5–8
74. Szabo N (1997) Formalizing and securing relationships on public networks. First Monday 2(9):2–7
75. Koulu R (2016) Blockchains and online dispute resolution: smart contracts as an alternative to enforcement. ScriptTED 13(1):67–69
76. Dmitry E, Pavel R (2018) The all-pervasiveness of the blockchain technology. Procedia Comput Sci 123(19):118
77. Qiqi G (2017) Cosmopolitanism and china in the age of artificial intelligence. Foreign Theoret Trends No 9:12–13 (in Chinese)
78. Daugherty P, Wilson J (2018) Human + machine: reimagining work in the age of AI. Harvard Business Review Press, Boston, pp 175–176
79. Kshetri N (2017) Blockchain's roles in strengthening cybersecurity and protecting privacy. Telecommun Policy 41(10):1027–1038
80. Vigna P, Casey M (2018) The truth machine: the blockchain and the future of everything. St. Martin's Press, London, pp 203–204
81. Makridakis S (2017) The forthcoming artificial intelligence (AI) revolution: its impact on society and firms. Futures 90(5):55
82. Qiqi G (2017) Global governance, mobility of people and community with a shared future for mankind. World Econ Politics No. 1:40 (in Chinese)
83. Fadilpašić S (2018) What six nobel laureate economists have to say about crypto. Cryptonews, March 31, 2018, https://cryptonews.com/exclusives/what-six-nobel-laureate-economists-have-to-say-about-crypto-1402.htm
84. Rooney K (2018) Bitcoin is the 'mother of all scams' and blockchain is most hyped tech ever. CNBC, October 11, 2018, https://www.cnbc.com/2018/10/11/roubini-bitcoin-is-mother-of-all-scams.html
85. Fry J, Cheah E-T (2016) Negative bubbles and shocks in cryptocurrency markets. Int Rev Financ Anal 47(32):343–352
86. Ciaian P, Rajcaniova M (2016) The digital agenda of virtual currencies: can BitCoin become a global currency? IseB 14(4):883–919
87. Bjerg O (2016) How is bitcoin money? Theory Cult Soc 33(1):53–72

88. Böhme R, Christin N, Edelman B, Moore T (2015) Bitcoin: economics, technology, and governance. J Econ Perspect 29(2):226–229
89. Qian Y (2018) Currency evolution logic under consensus mechanism and digital fiat currency issuance based on artificial intelligence. J Financial Studies 9(1):47. (in Chinese)
90. Ciaian P, Rajcaniova M (2016) The digital agenda of virtual currencies: can BitCoin become a global currency? IseB 14(4):35–36
91. Elbeck M, Baek C (2015) Bitcoins as an investment or speculative vehicle? A first look. Appl Econ Lett 22(1):33–34
92. Bryans D (2013) Bitcoin and money laundering: mining for an effective solution. Indiana Law J 1(89):472
93. Casey M, Vigna P (2018) In Blockchain We Trust. MIT Technol Rev, April 9, 2018, https://www.technologyreview.com/s/610781/in-blockchain-we-trust/
94. Furlonger D, Valdes R (2016) Hype cycle for Blockchain technologies and the programmable economy. Gartner, July 27, 2016, https://www.gartner.com/doc/3392717/hype-cycle-blockchain-technologies-programmable
95. Scott B (2016) How can cryptocurrency and Blockchain technology play a role in building social and solidarity finance? Working paper of the united nations research Institute for Social Development, pp 13–16
96. Akerlof G (1970) The market for 'lemons': the quality of uncertainty and the market mechanism. Quart J Econ 84(3):488–499
97. Manski S (2017) Building the blockchain: the co-construction of a global commonwealth to move beyond the crises of global capitalism. Strateg Chang 26(5):517
98. Abeyratne S, Monfared R (2016) Blockchain ready manufacturing supply chain using distributed ledger. Int J Res Eng Technol 5(9):2–3
99. Hayek F (2000) translated by Deng Zhenglai, Law, legislation and freedom. Encyclopedia of China Press, Beijing, p 54. (in Chinese)
100. Nikolakis W, John L, Krishnan H (2018) How Blockchain can shape sustainable global value Chains: an evidence, verifiability, and enforceability (EVE) framework. Sustainability 10(3926):13–15
101. Lemieux V (2016) Trusting records: is Blockchain technology the answer? Rec Manag J 26(2):137–138
102. Luca M (2017) Designing online marketplaces: trust and reputation mechanisms. Innov Policy Econ 17(1):89
103. Swan M (2015) Blockchain thinking: the brain as a decentralized autonomous corporation. IEEE Technol Soc Mag 34(4):49–51
104. Coeckelbergh M, Reijers W (2016) Cryptocurrencies as narrative technologies. Acm Sigcas Comput Soc 45(3):176–177
105. Kshetri N (2017) Can Blockchain strengthen the internet of things? IT Professional 19(4):68–69
106. Jackson R (2018) Why Blockchain B2B will be the megatrend of 2019. Big Think, 13 December, 2018, https://bigthink.com/technology-innovation/blockchain-b2b-is-the-future
107. Xiaoling W, Ge W (2014) 'New wyatt plan' or 'new Keynesian plan': how to construct a stable and effective international monetary system. Exploration Free Views, No. 8, pp 59–62. (in Chinese)
108. Gardner R, Diplomacy S-D (1956) Anglo-American collaboration in the reconstruction of multilateral trade. Clarendon Press, Oxford, pp 71–80
109. Hayek F (1990) Denationalisation of money: the argument refined. The Institute of Economic Affairs, Wesrminster, pp 23–56
110. Mundell R (1961) A theory of optimum currency areas. Amer Econ Rev 51(4):657–665
111. Mckinnon R (1963) Optimum currency areas. Amer Econ Rev 53(4):717–725
112. Kenen P (1969) The theory of optimum currency areas: an eclectic view. In: Mundell R, Swoboda A (eds) Monetary problems of the international. The University of Chicago Press, Chicago, pp 40–41
113. Ingram J (1959) State and regional payments mechanisms. Quart J Econ 73(4):619–632

114. Cooper R (1984) A monetary system for the future. Foreign Aff 63(1):166–167
115. Alessandrini P, Fratianni M (2010) Dominant currencies, special drawing rights and Supernational bank money. World Econ 10(4):62–63
116. Ocampo J (2010) Building an SDR-based global reserve system. J Globalization Dev 1(2):20–21
117. Xiaochuan Z (2009) Reflections on reforming the international monetary system. China Finance No. 7:8–9 (in Chinese)
118. Aiyar S (2009) An international monetary fund currency to rival the dollar? Why special drawing rights can't play that role. Cato Development Briefing Paper, No. 4, pp 6–8
119. Aiyar S (2010) Crisis prevention through global surveillance: a task beyond the IMF. Cato J 30(3):494–495
120. Aiyar S (2009) Can IMF currency replace the dollar? Times of India, April 5, 2009, https://timesofindia.indiatimes.com/sa-aiyar/swaminomics/Can-IMF-currency-replace-the-dollar/articleshow/4360430.cms
121. Triffin R (1960) Gold and the dollar crisis. Challenge 9(2):40–43
122. Eichengreen B (2009) Commercialize the SDR. Project Syndicate, April 27, 2009, https://www.project-syndicate.org/commentary/commercialize-the-sdr
123. Stei B (2007) The end of national currency. Foreign Affairs 86(3):95
124. Giaglis G, Kypriotaki K (2014) Towards an Agenda for information systems research on digital currencies and bitcoin. Bus Inf Syst Workshops, 3–13
125. Ali R, Barrdear J, Clews R, Southgate J (2014) Innovations in payment technologies and the emergence of digital currencies. Bank England Quart Bullet 54(3):262–275
126. Bech M, Garratt R (2017) Central bank cryptocurrencies. BIS Quart Rev 1(6):55–70
127. Fung B, Halaburda H (2017) Central bank digital currencies: a framework for assessing why and how. Discussion Papers from Bank of Canada
128. Masciandaro D (2018) Central bank digital cash and cryptocurrencies: insights from a new Baumol–friedman demand for money. Australian Econ Rev 51(4):540–550
129. Qian Y, Tang Y (2017) Some thoughts on central bank-issued digital currency. J Financial Res 7(7):78. (in Chinese)
130. Fanning K, Centers D (2016) Blockchain and its coming impact on financial services. J Corporate Account Finance 27(5):55–56
131. Reijers W, Coeckelbergh M (2018) The Blockchain as a narrative technology: investigating the social ontology and normative configurations of cryptocurrencies. Philos Technol 31(1):127–128
132. Yeomans M (1999) The quest for a global E-currency. CNN, September 28, 1999, http://edition.cnn.com/TECH/computing/9909/28/global.e.currency.idg/index.html.
133. Doguet J (2013) The nature of the form: legal and regulatory issues surrounding the bitcoin digital currency system. Louisiana Law Rev 73(4):1128–1131
134. Grinberg R (2012) Bitcoin: an innovative alternative digital currency. Hastings Sci Technol J 4(1):206–207
135. Xiangguang C, Zeqing H (2018) The formation mechanism of currency anchors and the maintenance of currency quality: collaterally study on the anchor of digital currency. J Renmin Univ China 32:92. (in Chinese)
136. Lei Z, Chengguo Z (2017) Research on the evolution of digital currency under the innovation of Blockchain technology: theory and framework. Economist No 5:81. (in Chinese)
137. Müller B, Elsman M, Henglein F, Ross O (2015) Automated execution of financial contracts on Blockchains. J Econ Perspect 29(2):233–234
138. Qian Y (2018) Currency evolution logic under consensus mechanism and digital fiat currency issuance based on artificial intelligence. J Financial Res 9:48. (in Chinese)
139. Rahman A (2018) Deflationary policy under digital and fiat currency competition. Res Econ 72(2):176–179
140. Vigna P, Casey M (2015) The age of cryptocurrency: how bitcoin and digital money are challenging the global economic order. St. Martin's Press, London, pp 3–4

141. Teramoto S, Jurcys P (2014) Intermediaries, trust and efficiency of communication: a social network perspective. In: Fenwick M, Van S, UytselStefan Wrbka, Networked Governance, Transnational Business and the Law, Heidelberg: Springer, Berlin, Heidelberg Press, pp 99–101
142. Masciandaro D, Central bank digital cash and cryptocurrencies: insights from a new Baumol-friedman demand for money. Australian Econ Rev 51(4):547–549
143. Mägerle J, Maurer D (2009) The continuous linked settlement foreign exchange settlement system (CLS). Swiss National Bank, https://www.snb.ch/en/mmr/reference/continuous_linked_settlement/source/continuous_linked_settlement.en.pdf
144. Kocherlakota N, Wallace N (2004) Incomplete record-keeping and optimal payment arrangements. J Econ Theory 81(2):287–288
145. Salah K, Suliman A, Husain Z, Abououf M, Alblooshi M (2019) Monetization of IoT data using smart contracts. IET Networks 8(1):37
146. Dwyer G (2015) The economics of bitcoin and similar private digital currencies. J Financ Stability 17(11):89–90
147. Jun M (2018) Blockchain government—a next form of infrastructure for the twenty-first century. J Open Innov Technol Market Complexity 4(1):7
148. Walch A (2018) Open-source operational risk: should public blockchains serve as financial market infrastructures?. In: Kuo Chuen DL, Deng R (eds) Handbook of blockchain, digital finance, and inclusion, Volume 2, ChinaTech, Mobile Security, and Distributed Ledger, New York: Academic Press, pp 243–269
149. Dierksmeier C, Seele P (2018) Cryptocurrencies and business ethics. J Bus Ethics 152(1):12–13
150. Wei Z, Wei W et al (2018) Digital currency: Fintech and currency reconstruction. CITIC Press, Beijing, pp 79–80. (in Chinese)
151. Dwyer G (2015) The economics of bitcoin and similar private digital currencies. J Financ Stability 17(10):86
152. Chaoyang W, Bugao Z (2015) Ripple in internet finance: principles, models and challenges. Shanghai Finance No. 3, pp 46–52. (in Chinese)
153. Mouradian A, Augeblum I, Valois F (2014) RTXP: a localized real-time mac-routing protocol for wireless sensor networks. Comput Netw 67(10):43–59
154. Armknecht F, Karame G, Mandal A, Youssef F, Zenner E (2015) Ripple: overview and outlook. Trust Trustworthy Comput, 163–180
155. Keju S (2016) A comparison research on ripple and SWIFT. Financ Theory Pract 10:105–107. (in Chinese)
156. Caplinger D (2018) Find out about this up-and-coming cryptocurrency: 5 facts you didn't know about ripple. Motley Fool, March 11, 2018, https://www.fool.com/investing/2018/03/11/5-facts-you-didnt-know-about-ripple.aspx
157. Deudney D, Ikenberry J (2018) Liberal world: the resilient order. Foreign Affairs 97(4):16
158. Pogge T (20120), translated by Liu Zui et al, Kant, Rawls and global justice. Shanghai Translation Publishing House, Shanghai, p 198. (in Chinese)
159. Yuyan Z (2016) A Chinese perspective on global governance. World Econ Politics 9(7). (in Chinese)
160. Zhicheng W (2016) The influence of global governance upon state governance. Soc Sci China 6(6):28. (in Chinese)
161. McKinnon R (1984) Currency substitution and instability in the world dollar standard. Amer Econ Rev 74(5):1129–1131
162. Daoping W, Xiaoyun F (2011) Is the current international monetary system the cause of global economic imbalances and financial crises? J World Econ 34:69. (in Chinese)
163. Tao G, Yuchao Z, Zheng G (2014) Unfinished reform: international monetary system in the post-bretton woods era. Studies Int Finance 10:6. (in Chinese)
164. Harvey D (2017) translated by Zhang Yin, The limits to capital. CITIC Press, Beijing, p 396. (in Chinese)

165. Yide Q, Mingqi X (2011) Strengthening the role of SDR in international monetary system. Int Econ Rev 3(3):55
166. Lei Z (2019) Research on digital credit co-governance model in the innovation of fintech. J Soc Sci 2:55. (in Chinese)
167. Jinhan L, Jian C (2018) Research on the influence of digital currency on international monetary system. Economists 5. (in Chinese)
168. Xiong Z (2004) The money illusion: interpreting the marxist philosophy of history. Soc Sci China No 4:54–55. (in Chinese)
169. Changhe S (2017) Order and governance in a connectivity world. World Econ Politics 2:27. (in Chinese)
170. Beitz C (2012) translated by Cong Zhanxiu, Political theory and international relations. Shanghai Translation Publishing House, Shanghai, p 116. (in Chinese)
171. Filippi P, Davidson S, Potts J (2018) Blockchains and the economic institutions of capitalism. J Inst Econ 14(4):640–641
172. Tapscott D, Tapscott A (2018) Blockchain revolution: how the technology behind bitcoin is changing money, business, and the world. Penguin Canada Press, pp 179–192
173. Levin R, Waltz P, LaCount H (2018) Betting Blockchain will change everything: SEC and CFTC regulation of blockchain technology. In: Kuo Chuen DL, Deng R (eds), Handbook of Blockchain, digital finance, and inclusion, Vol 2, ChinaTech, Mobile security, and distributed ledger, New York: Academic Press, pp 209–210
174. Pazaitis A, Filippi P, Kostakis V (2017) Blockchain and value systems in the sharing economy: the illustrative case of Backfeed. Technol Forecast Soc Change 125(10):105–106
175. Dodgson M, Gann D, Berger I, Sultan N (2015) Managing digital money. Acad Manag J 58(2):330–332
176. Hongyu L, Xiaosan L (2012) International monetary power and geopolitical conflict. J Univ Int Relat 1:11
177. Ping C, Qingyou G (2011) The currency dimension of great-power games: Franco-American currency confrontation in 1960s and its historical revelations. World Econ Politics 4:45. (in Chinese)
178. Wei L (2016) Partners, institutions and international currencies: the international political foundations for the rise of the Renminbi. Soc Sci China 5:91
179. Fratianni M (2012) The future international monetary system: dominant currencies or supranational money? an introduction. Open Econ Rev 23(1):10
180. Helleiner E, Kirshner J (2009) The future of the dollar: whither the key currency? In Helleiner E, Kirshner J (eds), The future of the dollar. Cornell University Press, New York, p 8
181. Haass R (2018) How a world order ends and what comes in its wake. Foreign Affairs 98(1):28–30
182. Xiao L (2011) Why has monetary cooperation in east asia suffered setbacks?—on the internationalization of RMB and its implication for future monetary cooperation in East Asia. Int Econ Rev 1:118. (in Chinese)
183. Guoyou S (2013) A study of U.S.-china financial relations. Cunrrent Affairs Press, Beijing, pp 66–67
184. Russet B (1985) The mysterious case of vanishing hegemony; or, is mark twain really dead? Int Organ 39(2):218
185. Guoyou S (2018) China, the united states and global economic governance. J Soc Sci 8:27–28. (in Chinese)
186. Xiao L, Yongqi F (2012) International monetary system reform: collective actions and the role of G20. World Econ Politics 2:123–124. (in Chinese)
187. Kindleberger C (2003) translated by Gao Zugui, World economic primacy (1500–1990), Beijing: Commercial Press, pp. 306–307. (in Chinese)
188. Snidal D (1985) The limits of hegemonic stability theory. Int Orga 39(4):579–614
189. Mingli L (2017) Dollar hegemony and European financial security. J Int Security Studies 35:105. (in Chinese)

190. Tuo C (2010) Defining the international role of contemporary china: some reflections. Soc Sci China 5:132
191. Jinping X (2017) Building a community of human destiny together—speech at UN Headquarters in Geneva. People's Daily, January 20, 2017. (in Chinese)
192. Qiqi G (2018) A community of shared future for humanity and world politics in the era of artificial intelligence. Contemp World Socialism 3:47. (in Chinese)
193. Ruguo F (2017) World risk society' governance: the paradigm of complexity and Chinese participation. Soc Sci China 2(2):82. (in Chinese)
194. Hileman G, Rauchs M (2017) Global cryptocurrency benchmarking study. Cambridge Centre for Alternative Finance, March 27, 2017, https://www.jbs.cam.ac.uk/fileadmin/user_upload/research/centres/alternative-finance/downloads/2017-global-cryptocurrency-benchmarking-study.pdf
195. Xuelian L, Lu Y (2016) The implications of state governance for effective global governance. Soc Sci China 6:33. (in Chinese)
196. Qiqi G (2018) "The 'risks of marginalization' faced by developing countries and the mission of china in the age of artificial intelligence. Int Rev 4:44. (in Chinese)
197. Wang H, Li L (2018) A brief analysis of digital stabilization tokens. China Finance 19:45–46
198. Castor A (2018) Gemini and Paxos Both Launch Stablecoins on Ethereum Blockchain. Bitcoin Magazine, September 10, 2018, https://bitcoinmagazine.com/articles/gemini-and-paxos-both-launch-stablecoins-ethereum-blockchain/
199. Galicki A (2018) U.S. Sanctions Venezuela's 'petro' cryptocurrency amid broader trend of sanctioned and rogue regimes experimenting with digital assets. CLEARY GOTTLIEB, April 13, 2018, https://www.clearyfintechupdate.com/2018/04/u-s-sanctions-venezuelas-petro-cryptocurrency-amid-broader-trend-sanctioned-rogue-regimes-experimenting-digital-assets/
200. Zhong X (2018) The new progress in the construction of CIPS. China Finance 11:41–42. (in Chinese)
201. Kuo Chuen DL (2015) Handbook of digital currency: bitcoin, innovation, financial instruments, and big data. J Wealth Manag 18(2):96–97
202. Yiran S (2013) The characteristics of global economic governance from the international system to the world system. Studies Int Relat 1:88. (in Chinese)
203. Cotton J (2019) Constructing sustainable world order: one century's global governance. Int Rev 1:69

Chapter 5
Blockchain, Global Governance, and China's Grand Strategy

Intelligence and blockchain technology are precisely the AB side of intelligent revolution technology. The current dilemma of the global governance system is mainly reflected in the ineffectiveness of the political system leading to international anarchy, the hegemonic position of developed countries in the economic system, the rise of emerging countries creating a sense of relative deprivation in developed countries, and the head-on collision between emerging countries and developed countries that will generate structural conflicts. The revolutionary significance of blockchain for the reform of the global governance system mainly lies in the following. Firstly, the financial funding system of the United Nations is constructed by issuing global digital currency. Second, realizing high-level democracy of the UN system is based on the chain of assets of each country.

5.1 The Future of Blockchain and the Intelligent Revolution

Essentially, blockchain is a distributed ledger, a multi-centric bookkeeping method.[1]Bookkeeping is an act of recording information related to economic activities. In human social actions, bookkeeping is a necessary power. Most of the previous bookkeeping behaviors show centralized characteristics. For example, as a new payment tool, Alipay is the center of the mobile payment business. Activities such as the flow of funds of the parties involved are recorded on Alipay's servers. Unlike the traditional centralized bookkeeping function, this blockchain function is based on cryptography.[2] Through two-way cryptographic activities, blockchain can keep the process of bookkeeping on different subjects and guarantee anonymity and security for the parties to the transaction.

---

[1] Kuo, Kim and Ohno-Machado [1], pp. 1211–1220.

[2] Sathya and Banik [2], pp. 279–288.

Q. Gao and J. Zhang, *Artificial Intelligence Governance and the Blockchain Revolution*, Artificial Intelligence and the Rule of Law,
https://doi.org/10.1007/978-981-99-9211-9_5

The emergence of blockchain technology has provided an essential technological basis for governance activities. In the last three decades, governance has become a central concept in social sciences. Governance emphasizes connotations such as multi-center, transparency, trust, and justice, and blockchain is precisely very close technically to the conditions for realizing these connotations. First, blockchain has a polycentric character, changing the previous single, centralized bookkeeping method. This new way of bookkeeping allows every subject to participate in bookkeeping activities, while each issue will have a separate ledger in their hands. Although many scholars have emphasized the decentralized character of blockchain during its dissemination, it is not decentralized but rather polycentric. Blockchain still has a center, which is no longer a single center but a multi-center. Scholars who emphasize decentralization often hope to promote the spread of blockchain ideas through exaggerated views. Second, blockchain has the feature of traceability. Since each participating entity has ledgers and all transaction activities are recorded in multiple ledgers, then all transaction activities can be confirmed by comparison between ledgers and ledgers. The feature of traceability can increase the transparency of transaction activities. Third, blockchain can increase trust. Due to the enhanced trustworthiness and accuracy of bookkeeping activities, people's transaction behaviors will be more accurately recorded, increasing confidence. And this trust is not simply based on social trust among people but on algorithm-based digital trust. Some literature discusses blockchain features with an emphasis on the de-trusting nature of blockchain. Blockchain does not de-trust but increases trust between people, except that this trust is secured and transmitted through numbers and algorithms.

The primary technological foundation of the intelligence revolution is artificial intelligence. What artificial intelligence enables is the intelligence of things. In short, the things we see before are static or mechanical. Through artificial intelligence technology, static things can move and realize the intuitive interaction between things through information transfer and processing. Artificial intelligence is a further development of the Internet and the Internet of Things technology. The Internet first realized the connection of terminals such as computers, while the mobile Internet realized the connection between cell phones. The Internet of Things further expands the subject and scope of connection, which can realize the interconnection and automatic communication among all things, thus generating new productivity. Artificial intelligence is the life-like or human-like transformation of things, the essence of which is to imitate the behavior of animals or people as well as thinking and other contents.

Artificial intelligence will face two difficulties in the next stage of development. The first is the issue of security. Along with the proliferation of intelligent bodies, the concern of human society is that if hackers carry out terrorist activities by attacking AI systems, then the damage this produces will be massive. For example, on future intelligent transportation systems, the damage would be enormous if hackers were to create crashes on the road by manipulating or interfering with driverless systems. Similarly, hacking attacks against critical infrastructure or intelligent medical devices are equally alarming. The traceability of blockchain can help solve such problems. The future trend will be to achieve a complete record of all network activities through blockchain technology. While the previous Internet traceability is still centralized, future blockchain-technology-supported traceability can prove the authenticity of Internet activities through a multi-center approach. Meanwhile, this two-way recording method is an open supervision mechanism. By recording all Internet activities, hackers' identities can be more accurately determined, and severe legal punishment and deterrence can be imposed on hackers' attacks.

Similarly, privacy protection is also a significant problem facing the development of artificial intelligence at present. Because of the emergence of many bits of intelligence, personal privacy can easily flow into public space, and some unscrupulous people will buy and sell personal privacy as property. Blockchain can retain original data and control data sharing through cryptography and cryptographic principles. Encryption technology can give certain rights to data users. According to different business needs, scenarios, and contents, users will be given other access keys and leave timestamps during the use process. Then the data usage will be easier to trace, and the essential data protection will be easier to be achieved.

The further development of AI in the future will generate more complex multiple social relationships, including the problem of interaction between AI agents and humans. The social relations in traditional society are mainly among humans, and only under certain circumstances will they involve the relations among humans, animals and nature. But new social relationship changes will arise with the development of technologies such as artificial intelligence and the Internet of Things. For example, there may be several or even dozens of intelligent bodies in a family, so if the interaction among these intelligent bodies and people still needs to be handled by people, then eventually, humans will be overwhelmed. The original purpose of intelligent bodies is to help humans solve problems. Still, if they generate many new complex relationships that require human effort to solve, they will become a burden to humans. In this sense, smart contracts become crucial. A smart contract is a transaction between intelligence that is automatically reached through some algorithm, according to some principles and guidelines of human society. There are two keys to this: first, the agents should get deals automatically among themselves, and second, the agents need to imitate some common cultural principles of human society. This must be dangerous if smart contracts are made against fundamental human values. Therefore, the design of smart contracts must be consistent with actual human deals and standard codes of conduct. In this sense, intelligence technology and blockchain technology are the AB side of the intelligent revolution technology.

First, China needs to conduct more in-depth research on the underlying technology. Most of the core applications of blockchain are currently driven by the West, Such as Bitcoin, Ethereum's Ether, IBM's Hyperledger Project, etc. These projects are significant in blockchain, yet Western companies or associations drive their underlying architectures. Most of the domestic blockchain companies make superficial application changes on the underlying architecture of the Western companies. In other words, the core systems of these critical infrastructures are still in the hands of the West. Key technologies are crucial in the current trade friction between the US and China. If blockchain is vital to the future development of China's innovative economy, then the localized replacement of such underlying technologies is critical. Therefore, China needs to make different layouts in the underlying technology.

Secondly, China can promote the implementation of some specific applications. For example, some local governments can be encouraged to conduct pilot national digital currencies, and enterprises (huge private enterprises such as Tencent and Ali) can be encouraged to do digital currency pilots. Such pilot work has already taken shape in the United States. For example, New York State officials have established a working group to study how to regulate, define and use cryptocurrencies properly. Recently, Facebook is also planning a new digital currency, Libra, and it can be seen that the U.S. is also taking a pilot approach to deal with blockchain. The pilot approach has been critical in driving reform in our country during the reform and opening-up process. Pilots have caused essential changes in many places. Since blockchain is vital to the future smart revolution, it is necessary to establish a regional pilot mechanism for blockchain.

Thirdly, China can gradually form its own technical and social standards in the piloting process. In the new smart revolution in the future, competition with leading Western countries in technology applications will be frequent. We have shown certain advantages in the current intelligence revolution process. Competition for standards will become essential in future competition because the standard means the institutional discourse and the future competition path and direction.

5.2 Blockchain's Revolutionary Implications for the Transformation of the Global Governance System

We need to think about changing the global governance system from the perspective of the future, not just from the present perspective, i.e., to do something similar to Kant's thinking. Kant could not be accepted by the time when he proposed the theory of perpetual peace when Europe was amid strife and war.[3] However, Kant proposed the doctrine of perpetual peace beyond the context and circumstances of his time and laid the foundation for the League of Nations and the practice of the United Nations a century later. Thus, the Kantian solution is a transcendental one. This metaphysical concept is needed for our current thinking about the change of the system of global governance.

The global governance system has two logics: explicit logic and implicit logic.[4] From the explicit logic, the earliest form of the global governance system is the League of Nations after World War I, but it is immature. The post-World War II United Nations system is a more mature global governance system, mainly including the political, economic, and social systems. The political system is the core system of the UN, primarily consisting of the Security Council and the UN General

---

[3] Kant [3], pp. 106–142.

[4] Qiqi [4], pp. 3–12.

Assembly. The economic system is the Bretton Woods system, which consists of GATT/WTO, IMF, World Bank, etc. The social system is the subsidiary system of the United Nations, mainly composed of the World Health Organization, UNESCO, the International Atomic Energy Agency, the International Telecommunication Union, etc.

In the early years of the UN's operation, the political system was more characterized by U.S. hegemonic control, typified by the U.S.-led "United Nations forces" in the Korean War. After that, the system slowly diversified, most notably through the Soviet Union's repeated use of the veto to curb U.S. hegemony. Another representative event was China's return to the UN in 1971. During the Ghali period, the political system of the UN became more diverse. In the economic system, the U.S. is dominant in the WTO, IMF, and WB, especially in the IMF. Even after the 2015 share adjustment, the U.S. still has 17.407% of the voting power in the IMF (International Monetary Fund), Japan has 6.464%, and China has 6.394%.

In addition to the explicit global governance system, an implicit global governance system exists. This system emerged at the international macro level after the Yom Kippur War in 1973. In response to the oil crisis and inflation, the developed Western countries formed a new club, the G7. Over time, the G7 gradually evolved into the most critical mechanism for the West in global governance. At the same time, a global governance system under the implicit hegemony of the United States exists in many fields, such as finance and information technology. In this process, several commercial organizations have played a crucial role. Although these commercial organizations claim to be private institutions, they all serve the U.S. hegemony in some sense. They are either under the direct influence of the U.S. government or are voluntarily guided by it. They include the Society for Worldwide Interbank Financial Telecommunication (SWIFT), the New York Clearing House Interbank Payment System (CHIPS), the Internet Corporation for Assigned Names and Numbers (ICANN), the Institute of Electrical and Electronics Engineers (IEEE), the International Wi-Fi Alliance (WFA), and others.

On the whole, the current dilemma of the global governance system is mainly reflected in the following aspects:

First, the weakness of the political system has led to global anarchy.[5] This is a long-term problem faced by the global governance system, and it is also a structural problem. The core of the global governance system is the UN, and the fundamental weakness of the core UN system is its need for sufficient power. The UN coordination mechanism needs to be stronger, and the real reason is that the UN has no tax system and no direct funding source. The UN has been financed mainly by contributions made by countries, resulting in an embarrassing situation in which the UN has to rely on individual sovereign states. Because there is no source of revenue, the UN has difficulty maintaining its own directly recruited or constituted military forces, depending instead on sovereign states to form UN peacekeeping forces. In a sense, the UN does not have a "people" that strongly identifies with it.

[5] Xi [5], pp. 3–9.

Second, the hegemonic position of developed countries in the economic system has perpetuated a sense of injustice in developing countries. In the global governance system, it is the G7 framework that has long played an important role. This framework is effectively a club for developed countries, and as a result, developing countries have long felt chronically excluded.

Third, the rise of emerging countries has created a sense of relative deprivation in developed countries. The United States often criticizes China as the biggest beneficiary of the WTO. But this criticism is a relative perception. The U.S. has come to this conclusion from a narrow perspective that China's economy has grown significantly in the 20 years since joining the WTO. Some leaders in the developed West have adopted unilateralist policies, such as "withdrawal," giving vent to this sense of relative deprivation. At the same time, emerging countries are demanding a new power structure. The new model of international economic cooperation represented by the G20 summit is a compromise solution under this reality.

Fourth, a head-on collision between emerging and developed countries can produce structural conflicts. This structural conflict has already manifested in the U.S.-China trade dispute. In the short term, there may be a trend of "de-globalization" of developed countries globally. Some developed countries in the West will take more actions, such as withdrawing from international or regional mechanisms.

Intelligence technologies include artificial intelligence and bio-intelligence (brain science, neuroscience, gene technology, etc.). The goal of Intelligence technologies is super intelligence, the essence of which is agglomeration technology. It can be expected that human productivity will be significantly improved based on Intelligence technologies. In a sense, if human beings make full use of Intelligence technologies, it can solve the problems of resource scarcity and labor shortage that human beings have been facing for a long time, and it can bring more products to human society and make a society of abundance a reality at the global level.

Blockchain comprises four powerful technologies: distributed ledger, peer-to-peer transmission, consensus mechanism, and cryptography. Regarding the degree of technical application: blockchain 1.0 is a digital currency, blockchain 2.0 is a smart contract, and blockchain 3.0 is an intelligent society. In other words, the more blockchain technology advances, the more it integrates with Intelligence technologies and constructs a smart society. The essence of blockchain technology is distribution, which will hopefully overcome the problem of centralization. Blockchain technology helps to reasonably address the privacy, security, and fairness issues that have intensified with the development of artificial intelligence. There are three forms of blockchain technology: first, public chain, whose core is decentralized, represented by Bitcoin; second, private chain, whose core is centralized, represented by Alipay or Q Coin; and third, consortium chain, whose core is multi-centric.[6]The consortium chain is the application scenario with practical promotion value. The decentralization of public chains is more of a labeled idea. Even among Bitcoin operations, there is a relatively centralized operations team because Bitcoin's development will encounter many problems, such as in recent years it involves expansion

[6] Singh, Kumar and Kathuria [6], pp. 269–288.

issues. Therefore, decentralization is just a signifying concept that Bitcoin hopes to produce. In practice, what is more feasible is a consortium chain that emphasizes multi-centralized.

Overall, the revolutionary significance of blockchain for the reform of the global governance system lies mainly in the following.

The first is to build a financial funding system for the UN by issuing a global digital currency. At present, it is difficult or almost impossible to make a complete tax system similar to that of a sovereign state at the UN level. Some scholars have also suggested that an aviation tax be levied to fund UN reform, etc., but these are only a drop in the bucket for UN operations. Hence, an essential breakthrough in the future is a currency, i.e., a global digital currency issued based on sovereign coordination among countries. In fact, since its inception, Bitcoin has wanted to fill such a role, but Bitcoin itself has significant limitations. Bitcoin is generated entirely outside national sovereignty, which is inherently unacceptable to sovereign nations. Moreover, Bitcoin is heavily used for black market transactions, money laundering, human trafficking, drug trafficking, and other criminal activities. Countries are gradually adopting a tightening policy towards Bitcoin. As governments gradually issue digital currencies, there will be less and less room for Bitcoin.

Given this, a global digital currency could be issued based on national coordination. There are two ideas: the first is to "transform the old stove," that is, to transform the SDR (Special Drawing Rights) of the International Monetary Fund into a digital currency that can be used as a medium of exchange, and the rights and interests of countries in this currency can follow the shares of countries in the previous SDR. The second idea is to "start a new one," i.e., to issue a new world currency based on blockchain technology. Again, this would be based on national coordination. Still, the shares would have to be re-negotiated, which could be based on population or economic aggregates, and involve coordination of other rules. In short, such a global digital currency is formed based on coordination among countries. After creating a global digital currency, governments could use cash to track various behaviors, especially criminal behaviors, such as telecom fraud, money laundering, etc. Therefore, criminal activities would be significantly reduced, while the space for Bitcoin would gradually disappear.[7] The United Nations could establish some new type of monetary organization or transform it into the International Monetary Fund by imposing a minting tax on the circulation and use of the world's currency worldwide. Due to the dominance of the U.S. dollar in international currency reserves, settlements, and payments, the U.S. currently receives an average of about $25 billion per year in minting taxes. This compares to the current UN fiscal outlay of approximately $2.5 billion annually. Therefore, this mint tax could significantly increase the economic capacity of the UN.

Second, the United Nations system should achieve a high level of democracy based on the chain of assets of each country. Based on the global digital currency, the critical support of each country should be "on-chain." This will effectively solve the problem of UN contribution arrears because it will be done based on smart contracts in the

[7] Ba, Zhang and Zhu Y [7], pp. 3–9.

future. Once the conditions are met, the issue of contribution arrears will be solved entirely. In addition, if some sovereign countries violate some activities, such as invading other countries or committing illegal acts, after the UN General Assembly and the Security Council discuss and make decisions in public, the punishment mechanism can be automatically implemented.

The critical significance of "on-chain" for global governance reform is that it can significantly enhance the effectiveness and enforcement of the UN mechanism. Based on the global digital currency, some of the assets of countries can be locked in the blockchain in the form of smart contracts. Once the smart contracts are triggered, the relevant agreements will automatically take effect. The United Nations will be able to coordinate more strongly and allocate resources when it is well-funded. A more hard-core global governance smart contract will be very effective for conflict resolution and can contribute to a fairness program in global governance. International mechanism is based on rules, and blockchain-related technologies can guarantee the enforcement of regulations, which can significantly save transaction costs. The distributed ledger in blockchain technology can realize the feature of joint supervision of multiple subjects in each country, which can effectively improve the transparency and mutual trust of countries' concerns about a particular issue. On the basis of the hard support of blockchain technology, the resources required for the operation of the United Nations can be guaranteed and a higher level of democracy can be achieved on its basis. The United Nations can apply the theory of representation to its practical application, such as forming a United Nations parliament, which can be promoted by referring to the development process of the European Parliament. With the addition of blockchain technology, this is perfectly feasible operationally.

5.3 Global Digital Currency Competition and China's Grand Strategy

A longer-term strategic conflict between China and the United States is essential to the future global order.[8] This conflict will unfold in five areas: trade, finance, science, technology, and culture. It is essential to see the long-term nature of the strategic competition between China and the United States. The U.S. will not tolerate the existence of a rival whose economic volume is almost equal to its own, so the U.S. will adopt various methods to hound China in these four areas, all of which it will have to face in the future. The trade conflict has evolved into the most direct surface conflict. Competition in finance has yet to unfold fully, and competition for digital currencies will be the focus of the financial sector in the future. Therefore, China sees this and is very cautious in issuing national digital currencies, etc., biding its time. In the technology sector, the U.S. will undoubtedly target a series of key core technology areas for China to restrict and block. Then China will need to break through these barriers through its technological efforts. Cultural competition is reflected more in the

[8] Tammen and Kugler [8], pp. 35–55.

ideological sphere, and the Hong Kong Issue 2019 reflects this cultural competition to some extent. From another perspective, after a long-term strategic conflict between China and the U.S., there will generally be a move toward détente. Negotiations at critical points and new consensus reached under new technological conditions will be more conducive to establishing a new global governance system. And at that time, blockchain and artificial intelligence technologies will be more mature, so it is likely to form a more effective global governance system.

The following discussion focuses on the future of global digital currency competition. In the future, there will be fierce competition between the DCEP (Digital Currency Electronic Payment) promoted by China and the virtual currency supported by the United States. China's push for DCEP stems from the central bank's particular research group on legal digital currencies, established in 2014. The group was renamed the Digital Currency Research Institute in 2016, and in 2018 Shenzhen Fintech, which serves the DCEP layout, was founded. On August 18, 2019, "Opinions of the Central Committee of the Communist Party of China and the State Council on Supporting Shenzhen in Building an Early Demonstration Zone of Socialism with Chinese Characteristics" specifically mentioned "supporting the development of digital currency research and innovative applications such as mobile payment in Shenzhen". China is confident in promoting DCEP because WeChat and Alipay already have a long history of practice and relevant results in China. For example, Alipay is already used in over 200 countries and regions as of 2018, supporting direct transactions in at least 20+ currencies, cross-border payments in 38 major countries and territories worldwide, and more. Alipay went live in 2004, while Apple Pay, a mobile payment tool in the United States, only went live in 2014, and Samsung Pay in South Korea in 2015.

DCEP is issued to protect China's monetary sovereignty and legal status and promote RMB internationalization. DCEP uses some blockchain technology but not peer-to-peer technology. Alipay and WeChat are corporate actions, while DCEP is a state action. Also, DCEP allows for dual offline payments. The issuance of DCEP can significantly strengthen the regulation of the financial sector. A series of criminal acts such as money laundering, gambling, telecommunication fraud, corruption, misappropriation of poverty alleviation funds, and receiving social insurance to buy luxury cars will be effectively restricted because the source and destination of every money can be effectively monitored based on DCEP.

Libra is a new type of digital currency to be released by Facebook, jointly managed by multiple major American companies. The reserve currency package in Libra's credit endorsement does not include RMB, and over 50% is in US dollars. If Libra receives support from the US government, there may be competition between DCEP and Libra in the future. The prospect of DCEP mainly depends on the following factors: first, whether China is resolute in issuing DCEP; second, whether the total economic volume of China can form a strong support for DCEP; third, the information technology foundation is an important guarantee; fourth, all kinds of assets need to be on the chain so that blockchain can not only stay on the concept.

In the future, with the aid of blockchain technology, China's global governance reform program could go in two directions:

First, the road of strength. "The Belt and Road" is strategically important to China's development until there is a structural de-escalation in U.S.-China relations. "The Belt and Road" is an open initiative in which DCEP can play an important role. In practice, Alipay and WeChat are already playing a similar position to commercial banks. Can a Belt Road Currency (BRC) mechanism be built in building the Digital Silk Road? Of course, such a new digital currency should be based on coordination of countries and the practice of China's DCEP. In short, The Belt and Road can be gradually institutionalized based on blockchain and artificial intelligence technologies.

Second, the road of virtue. The global mechanism will restart after the US and China truly move toward long-term détente and engage in dialogue. At this time, artificial intelligence and blockchain will play an essential role in it and can promote the reform of the core system of the United Nations. The institutionalized thinking in artificial intelligence and blockchain technology can be fully applied in the innovation of global governance system reform, and the reform of the core system of the United Nations will be the focus in the future. From the current perspective, China still emphasizes the concept of state-centered global governance. China's current focus is on national governance first and international governance second. The main conflicts we have to deal with are still self-centered, such as domestic security and stability issues. This self centered approach needs to be dominant for a long time and should not be challenged or questioned. But as China's total power continues to rise and the importance of its overseas interests comes to the fore, especially in the next two or three decades, China will need to put more emphasis on the role of the United Nations.

An increasingly assertive China will not follow the U.S. lead in acting as the world's policeman regarding international affairs. China's cultural practices and diplomatic habits make it uncomfortable with the idea of world policing. There is a profound religious element behind the U.S. approach to world policing. The U.S. participates in international interactions with a Christian mission spirit, using the world police approach to practice the Christian mindset of "light over darkness." This logic of hegemony is a Christian conversion mindset, i.e., the conversion of all non-Christians to Christianity.

However, Chinese cosmopolitanism is a communal cosmopolitan mindset, a notion that is transmitted from the inside out and in layers.[9] In other words, the Chinese need for the world originates from within. The Chinese ideal of the world is inside-out, so Chinese cosmopolitanism is a desire to extend Chinese governance practices outward, not in the form of a rigid extension, but on the premise that other countries will do so voluntarily. The way China sees it, imposing its will on the world is a meaningless hegemonic logic. Deng Xiaoping clarified China's aversion and opposition to hegemonic logic as early as the 1974 UNGA meeting. Thus, as China's overseas interests increase, China is bound to advocate a more vital role for the UN. However, strengthening the part of the UN does not mean giving the UN

[9] Qiqi [9], pp. 25–39.

Fig. 5.1 The Significance of Blockchain in Changing the Global Governance System

a status similar to that of a sovereign state but instead promoting the realization of power of the UN.

China has many contributions to make in achieving this goal. First, it can further improve and perfect the relevant system based on blockchain technology. Second, China can take the lead in issuing legal digital currency. Through the practice of digital currency in China, it can eventually introduce China's technical foundation and practical experience. This process can be promoted simultaneously at the UN's level and each country's level, so a systemic institution change centered on the blockchain can occur globally (Fig. 5.1).

References

1. Kuo T-T, Kim H-E, Ohno-Machado L (2017) Blockchain distributed ledger technologies for biomedical and health care applications. J Am Med Inform Assoc 24(6):1211–1220
2. Sathya AR, Banik BG (2020) A comprehensive study of blockchain services: future of cryptography. Int J Adv Comput Sci Appl 11(10):279–288
3. Kant I (1903) Perpetual peace (trans: Carpbell Smith M). London: George Allen and Unwin, pp 106–142
4. Qiqi G (2015) Theory and indexation of state participation in global governance. J Soc Sci 1:3–12 (in Chinese)
5. Xi L (2004) Crisis of international law. Law Rev 1:3–9 (in Chinese)
6. Singh S, Kumar A, Kathuria M (2012) Understanding the public, private and consortium consensus algorithms in blockchain technology. Int J Blockchains Cryptocurrencies 3(3):269–288
7. Ba S, Zhang D, Zhu Y (2020) The development and trend of global digital currency. J Financ Dev Res 11:3–9 (in Chinese)
8. Tammen RL, Kugler J (2006) Power transition and China-US Conflicts. Chin J Int Polit 1(1):35–55
9. Qiqi G (2016) Communal cosmopolitanism: an analytical framework for the interaction between global governance and state governance. World Econ Polit 11:25–39 (in Chinese)

Chapter 6
Intelligent Civilization and Intelligent Social Science

Intelligent science will need to be a separate discipline in the future, and it consists of three primary fields: intelligent science theory, intelligent engineering, and intelligent social science. Intelligent social science has three kernels: ethics, law, and policy. China's promotion of the construction of intelligent social sciences has two important meanings: firstly, it is an important breakthrough in building institutional discourse power, and secondly, it is an important guarantee for winning the competition of intelligent revolution. Forming an academic community requires four elements: an explicit research ontology, neat and open topics, mature research methods, and disciplinary boundaries with a certain degree of openness. Intelligent social science can be developed in the context of the development of new liberal arts.

6.1 The Necessity, Ontology and Methodology of Intelligent Social Science

In the context of the intelligent revolution, the development of intelligent social science is imperative. This section attempts to answer three questions: Why must intelligent social science be developed? What is intelligent social science? How can intelligent social science be constructed and developed? The author tries to offer his insights into these three challenging questions.

In the context of the successful development of artificial intelligence, intelligent social science has come into being. Intelligent social science not only involves the construction of a new, complex, and broad discipline but also relates to the significant efforts of the Chinese social science community to fight for international discourse. For now, there are two critical points in promoting the construction of intelligent social science in China:

First, it is an essential breakthrough for constructing institutional discourse. It is undeniable that Western countries led by the United States are at the forefront of

Q. Gao and J. Zhang, *Artificial Intelligence Governance and the Blockchain Revolution*, Artificial Intelligence and the Rule of Law,
https://doi.org/10.1007/978-981-99-9211-9_6

the world in legal norms, science and technology, and discipline construction. For example, in artificial intelligence, the leading and dominance of the West (especially the United States) are still evident. Not only in terms of technology but also in terms of relevant legal norms, ethical concepts, and innovative talents, the West is still in a dominant position. As far as the Chinese social science community is concerned, the current knowledge system is mainly formed under Western discourse. The classification regarding disciplines is copied from the West, and few original contributions are from Chinese academics. Since the twenty-first century, China's economic strength has increased significantly, and its influence in the international arena has grown, so it is increasingly important to break through the institutional discourse barriers of the West and develop a discourse system with Chinese characteristics. There is no doubt that discipline building will become an essential part of building institutional discourse in China in the future. How to create a new disciplinary system and knowledge system in some new spaces has become a crucial direction for contemporary Chinese scholars to work together. The author believes that governance science and intelligent social science will become two breakthroughs critical to the future Chinese social science community to build a discourse system with Chinese characteristics together.

Management science is a product of the industrialization era. Based on the first and second industrial revolutions, management science in the West came into being. In the 1980s and 1990s, the concept of governance emerged, and then governance became an essential concept in various disciplines. Governance and management are only one word apart in Chinese, but there is a clear difference in connotation. Management emphasizes top-down, strong constraints on people, while governance emphasizes two-way interaction and soft restrictions on people. Compared with management, governance has a stronger multi-centric connotation; therefore, governance science can combine new research topics and directions under interdisciplinary communication and become an important starting point for constructing new disciplines. In addition to governance science, intelligent social science is another vital breakthrough point for new fields in the future.

Secondly, it is an important guarantee to win the competition of intelligent revolution. The third industrial revolution is manifested as the information revolution, while the fourth industrial revolution is displayed as the intelligence revolution. Intelligent social science is a natural product of the intersection of the intelligence revolution with humanities and social sciences in terms of knowledge systems. Further, intelligent social science can be the "surprise" of the fourth industrial revolution to China's humanities and social sciences. For the first time in more than two hundred years, China is standing at the starting point of the new industrial revolution with the developed countries in the West. Therefore, China needs to seize the opportunity to fight for the initiative of the fourth industrial revolution, which requires contemporary Chinese social science scholars to join hands to build intelligent social science to guarantee the smooth and healthy development of the intelligent revolution. In short, intelligent social science has a leading significance for the intelligence revolution.

Any mature discipline must have an ontology. Ontology constitutes the essential kernel of the field. For example, the ontology of political science research is political phenomena. So, as an emerging discipline, what is the ontology of intelligent social science? The author believes that the ontology of intelligent social science should be a comprehensive ontology consisting of three levels of content: micro-ontology, meso-ontology, and macro-ontology.

At the micro level, the ontology of intelligent social science consists of both intelligences and people. As a product of the intelligence revolution, the intelligent body will become the emerging individual in the future social life, so the individual status of the intelligent body and the legal, social, and political relations attached to it becomes crucial. The human being is also the micro-ontology of intelligent social science because the intelligent body appears first as a human auxiliary. Still, at the same time, the important characteristic of the intelligent body is human-like. Therefore, whether to give human-like status to intelligence in the future will be a crucial issue in the study of intelligence. Additionally, this change in social relations will significantly impact people, for example, how the emergence of a large number of intelligence will affect labor and employment, so people affected under intelligence should also be an essential ontology for intelligence social science research.

At the meso level, the ontology of intelligent social science should be governmental, corporate, and social organizations that use human–machine interaction or collaboration. In the future, a large number of organizations will be mixed with human–machine. Therefore, the impact of mixed human–machine organizations on the organizational structure of future society, the reconstructive role of such organizational forms on economic activities and governmental governance processes, etc., will be the critical questions in the study of meso-ontology.

At the macro level, the ontology of intelligent social science is such great content as the human–machine world or intelligent civilization. The human–machine world can be an ontological description of future society. At the same time, intelligent civilization can take this description to a higher level, the same as previous agricultural, industrial, and information civilizations. The ontological topics addressed at the holistic macro level include the holistic challenge of general artificial intelligence to human beings, whether humans and machines can live together in harmony, and whether intelligent civilization represents social progress or social regression. Additionally, the maturity of a discipline is also demonstrated by the fact that it has a complete set of research methods. The philosopher Paul Feyerabend argued against research methods, arguing that a plan without a technique is the best. The strength of the research method is that it establishes boundaries for a discipline, i.e., not just anyone can talk about the discipline's content. Setting boundaries allows for a more focused and specialized discussion of the field. Therefore, research methods are essential for establishing limitations in intelligent social science, i.e., they are indispensable. However, because intelligent social science is a more open and futuristic discipline, its research methods should also be more inclusive. Overall, intelligent social science research methodology should contain at least three aspects.

First, the scientific method. Here, it includes the experimental and statistical methods, which are most strongly colored by natural and social sciences, and qualitative comparative analysis (QCA) or content analysis, which is relatively strong in humanistic features. The scientific method emphasizes scientific designs, such as formulating hypotheses, verifying laws, and reproducibility. New forms of artificial intelligence can also be introduced here, such as convolutional neural networks (CNN), recurrent neural networks (RNN), extended and short-time memory models (LSTM), generative adversarial networks (GAN), support vector machines (SVM), and knowledge graphs (KGP). The use of these methods will help generate many discoveries.

Second, the philosophical approach. Because much of what is explored in intelligent social science is entirely future-oriented, philosophical reflection is essential. The essential characteristic of human intelligence is learning by example, i.e., making a broader inference based on small data. Because philosophical reflection emphasizes the systematic structure of knowledge and its logical self-reference, it allows us to think more comprehensively about the emerging discipline of intelligent social science. It helps us to formulate some crucial hypotheses initially.

Third, the humanistic approach. Because many intelligent social sciences research objects occur in the future, specific research results are difficult to test by scientific methods. The philosophical approach emphasizes logical reasoning and intellectual structure, with higher demands on the researcher. In other words, not all research is amenable to the scientific method, and not all people can master the philosophical method. Thus, methods from the humanities, such as imagination, which is highly emphasized in works of science fiction, can also be a research method for intelligent social science. A classic example is that Asimov's "Three Laws of Robotics," a work of fiction, was later introduced into machine ethics and became the starting point for studying machine ethics. This is an excellent example of the critical role of a humanistic approach in intelligent social science research (Fig. 6.1).

Finally, we must return to Feyerabend's view of the "opposition to the method." Feyerabend's significance lies in proposing a new way of thinking, emphasizing that no definite method exists. In short, a practice-driven strategy should be adopted in intelligent social science research. Specific research designs and operational procedures should be accepted as long as the analysis is logical and convincing. In this way, research methods in intelligent social science can have boundaries while remaining sufficiently open-ended.

Fig. 6.1 The need, composition and research methods of intelligent social science

6.2 Disciplinary Composition, Issues, and Construction Path of Intelligent Social Science

Along with the deepening of the intelligence revolution, the author believes that, given its particular importance, intelligence science needs to become a separate discipline in the future, which consists of three primary disciplines: intelligence science theory, intelligence engineering, and intelligence social science. The idea of intelligent science focuses on the scientific approach centered on intelligence, such as the scientific theory related to artificial intelligence, biological intelligence, and superintelligence. Intelligent engineering is an engineering technology that applies innovative technology as a problem-solving method in various fields of society, which mainly includes concrete and operational engineering implementation paths such as machine learning and knowledge engineering.

Intelligent social science has three kernels: ethics, law, and policy. Ethics is the fundamental principle. The development of artificial intelligence must first pass the ethical hurdle. In terms of disciplines, ethics is more of a philosophical category. But in practice, ethics involves different application scenarios of other fields. Law is the norm for the development of artificial intelligence. The law mainly gives orders and prohibitions to tell people what they should and should not do. The security, privacy, and fairness issues AI will face in its subsequent development must be set down as binding norms. The third kernel is policy, i.e., the government's public guiding behavior in the field of AI, which plays more of a role in industry promotion and application promotion for AI development. In addition to these three kernels, the content of intelligent social science also radiates into various fields such as philosophy, political science, law, economics, sociology, and education. The composition of intelligent social science broadly reflects the characteristics of "3+X".

Intelligent social science is the intersection of intelligent technology and social science. The First Intelligent Social Science Forum, held at East China University of Political Science and Law from December 1 to 2, 2019, is a fundamental feature of intelligent social science characterized by interdisciplinary communication. Taking the keynote speaker as an example, Academician Chu Junhao of the Chinese Academy of Sciences launched a macro discussion from the perspectives of technical characteristics of artificial intelligence, macro trends, and its social impact. The keynote speaker has a high proportion of intelligent philosophy. For example, researcher Duan Weiwen from the Institute of Philosophy, Chinese Academy of Social Sciences, discussed how to use AI well and conduct intelligent reconfiguration of future society. Professor Wang Tianen from the School of Social Sciences of Shanghai University systematically analyzed the complete 29connotation of AI to reconfigure future culture by integrating science and philosophy. Professor Cheng Sumei, Deputy Director of the Institute of Philosophy, Shanghai Academy of Social Sciences, discussed the overall impact of AI on the future society from the macro perspective of intelligent civilization.

There can be two paths for the construction of an intelligent society: one is top-level design, and the other is spontaneous order. Top-level design is to plan and promote the construction and development of intelligent social science as a whole, with government guidance as the core. The spontaneous order is to gradually form a consensus through the joint discussion of the intellectual social science community. The advantage of the top-level design is that it advances quickly and has a specific holistic nature. This approach is most effective when implementing a catch-up strategy, i.e., one can learn quickly by referring to a template; however, when implementing innovative activities, one faces many difficulties. The construction of intelligent social science is a highly creative activity. The role of spontaneous order paths in constructing this discipline becomes even more critical as it is challenging to grasp the future face and direction of intelligent social science even at a high strategic level of government.

In the initial establishment, such an academic community is likely to be a "heroes' meeting," i.e., solving practical problems through interdisciplinary exchanges, or a "hodgepodge," i.e., simply a gathering of scholars from different disciplines, even if the academic language used by all of them is quite different. However, such a direct, face-to-face dialogue is often helpful in thinking holistically about essential and immediate issues, such as the ethics and policy of drones. We have to think about the solutions to these problems for practical reasons. In the interactive discussion of these problems, a consensus may gradually emerge or reach among the various disciplines. Then the "hodgepodge" will gradually become a "heroes' meeting." In the beginning, the study of intelligent social science is interdisciplinary, i.e., we need to step out of our respective disciplines or use the methods, tools, theories, thinking, and discourse of our respective fields to discuss issues related to artificial intelligence. The new intelligent social science will gradually become a unique academic community in gradual development.

The formation of an academic community requires the following four elements: first, an explicit research ontology; second, topics of common interest that are both neat and open at the same time; and third, a mature research methodology (mature research methods can keep outsiders out of the discipline); fourth, certain disciplinary boundaries while maintaining a certain openness to outsiders. Intelligent social science is still a system of knowledge in its formative years. At first, it should be more open, and the boundaries should be blurred or unnecessary. After some time, edges should gradually form, which is one of the signs of maturity in the discipline. Only some people can come and talk about issues related to intelligent social science. For discussions about brilliant social science to become more and more specialized and to have a higher technical content, it requires a continuous accumulation of knowledge within the discipline.

The future development of intelligent social science will focus on institutionalizing the content of discussions into outcomes and then gradually developing "shared knowledge," i.e., standard concepts, paradigms, and core ideas, in the results through the interaction of academics. In some large AI-related conferences organized by society or government, a joint manifesto is often standard, as it is an obvious form of presentation. However, the academic community needs manifestos and more the condition of solid, persuasive scholarship. An essential feature of artificial intelligence is that it can act as a "city of wisdom." A single intelligence unit is insignificant, but combining them for parallel computation can lead to enormous productivity. Again, it is essential to put this connotation into the process of intelligent social science formation.

Therefore, a vital task for us is to institutionalize the "Intelligent Social Science Forum" and gradually form a regular organization and regularize the activities. We will fix the relevant academic discussions intellectually, gradually provide some reference opinions for social development, and progressively take root in social science. Therefore, the construction of intelligent social science will need to be based more on spontaneous order rather than top-level design. Because no gifted scholar or government official will know the composition and essentials of future intelligent social science, we need more colleagues to work together within the community to form common concepts, methods, and discourses. In short, the construction of intelligent social science will be a process of spontaneous order formation.

6.3 The New Liberal Arts: A New Direction for Liberal Arts Education in the New Era

In recent years, the discussion on "new engineering" has been heating up. At one time, "new engineering" has become a core term in higher education and has received wide attention. However, the author believes that Chinese higher education should build "new engineering" and "new liberal arts."

Why do we need to build a "new liberal arts"? The reasons are as follows:

First, implementing national strategies requires not only a large number of new engineering and scientific talents but also many new liberal arts professionals. The country is implementing "Internet+," "Network Power," and other significant strategies, all of which as system projects, need the collaboration and support of the whole society. These strategies are systemic projects that require the cooperation and support of the entire community. For example, automakers sees 2020 as a critical year for driverless cars to be implemented, and the biggest obstacle to the implementation of driverless cars is not the technology itself but the relevant laws, public policies, and social psychology that go along with driverless cars. Therefore, implementing the national strategy requires many new liberal arts talents who understand the progress of technology and can actively respond to it.

Second, China needs to further strengthen its institutional discourse in the international community in the reform of liberal arts education. China has made outstanding achievements in the scale of liberal arts education. Still, the current social science textbooks are almost all dominated by Western knowledge, with much Western ideological content mixed in. In other words, although China is a major liberal arts country, it is not a leading liberal arts country. Chinese culture and knowledge need to be further theorized and disciplined, and their influence in the international community needs to be further enhanced. The new round of technological revolution has emerged a large number of emerging fields and propositions, such as the vast disruptive impact of technologies such as artificial intelligence, blockchain, genetic engineering, and virtual reality on human society and their prevention and control, which allow China to redefine liberal arts education.

What is the "new liberal arts"? The "new liberal arts" is building a liberal arts discipline that actively responds to technological and social changes, uses artificial intelligence and big data, and provides essential support for future social development and industrial upgrading in the context of a new round of technological changes. Specifically, the "new" focus of the "new liberal arts" is reflected in the following points:

First, broad crossover. The "new liberal arts" should be the product of the active combination of liberal arts and science and engineering knowledge. Under the influence of the new technological revolution, the traditional research methods of liberal arts will be significantly challenged. Previously, due to the high cost of data collection, social science research was mainly based on small data methods centered on sampling. In the context of the era of big data, on the other hand, social sciences need to analyze and process total samples of data, something that traditional liberal arts education does not provide. Therefore, in the future, liberal arts education will also need to learn programming languages such as Python, and liberal arts research will also need to master the techniques of extensive data collection and processing and use artificial intelligence techniques such as machine learning or knowledge mapping to analyze social science problems.

Second, future orientation. A new liberal arts education should cultivate talents who respond positively to future technologies and social changes. The contemporary liberal arts should be more concerned about the possible social impacts of technologies such as big data, artificial intelligence, and genetic engineering in the future. For example, the landing of artificial intelligence may have a significant impact on professions such as translators, clerks, junior lawyers, bank tellers, and car drivers in the future, and the talents cultivated by the new liberal arts need to have a more comprehensive understanding and judgment of these issues. For example, the development of the Internet of Things and artificial intelligence will make sensors ubiquitous, so how to regulate the risk of privacy leakage and security risks implied by this is also an issue of concern to the new liberal arts education.

Third, competence development. Traditionally, liberal arts education has been primarily vocational in nature, i.e., the major chosen is closely related to the career that follows. However, this career function model will be significantly challenged in the future. In recent years, banks have drastically reduced the recruitment of financial professionals and have all shifted to recruiting fintech, artificial intelligence, and blockchain talents. As a result, the future of liberal arts education should change to multifaceted competency development and interest-focused education. "Teaching people to fish is better than giving them fishes." Otherwise, in the face of drastic changes in the social environment, liberal arts students will face the dilemma of "unemployment upon graduation."

How should we promote the "new liberal arts" education in the future?

Firstly, comprehensive research and consensus building. The proposal of "New Engineering" has gone through the process of "Fudan Consensus," "Tianda Action," and "Beijing Guide." In this process, the Ministry of Education has organized all parties to conduct in-depth research and further build consensus among universities and society through academic seminars and other activities. The promotion of the "New Liberal Arts" also requires this process.

Second, it should be implemented in a categorized manner among different institutions and majors. Some leading units can be selected from comprehensive universities, comprehensive liberal arts universities, and universities with advantages in single disciplines, and exploration can also be carried out according to the disciplines of philosophy, law, economics, political science, sociology, education, etc. to find new growth points and breakthrough points for each major in liberal arts.

Third, the formation of emerging liberal arts cross-major. Vigorously develop emerging liberal arts majors related to new technologies such as big data, artificial intelligence, virtual reality, and genetic engineering. Of course, these explorations can be partially launched in the short term. Since artificial intelligence has a more significant impact on human society, we can promote the construction of various disciplines of intelligent social science, such as competition law, intelligent political science, intelligent sociology, intelligent ethics, intelligent journalism, and intelligent education first.

Fourth, update the knowledge cultivation system of liberal arts talents. We should introduce the latest development of society and technology and the latest industry requirements for talent cultivation into the teaching process and actively cultivate the innovative, intelligent, digital, and engineering thinking of liberal arts students, and enhance their comprehensive abilities in various aspects such as innovation and entrepreneurship, interdisciplinary crossover, independent lifelong learning, and communication and negotiation.

Fifth, to form several demonstration results. Universities of all kinds should review the situation, anticipate the problem ahead of time, and make bold reforms and early experiments according to the orientation and advantageous features of the universities to achieve critical breakthroughs in the following areas: building several new high-level liberal arts universities; building several emerging liberal arts majors urgently needed by society; building several new courses reflecting the latest development of culture and technology; training several high-level professional teachers who actively respond to social changes; building a series of interdisciplinary technology and social application R&D platforms.

Postscript

This is my third book on artificial intelligence. In the first book, *Artificial Intelligence: Taming the Sciviathan*, The question I hope to answer is, what is the overall impact of AI on future society? In *Artificial Intelligence II: Towards a Scitopia*, I hope to unravel the "Musk paradox" in the Western conception of AI. This paradox has been very influential in the Western AI academic and industrial communities. If this paradox is not solved philosophically, AI cannot be truly developed.

In this book, *Artificial Intelligence Governance and the Blockchain Revolution*, the question I hope to answer is, how do AI and blockchain actually relate to each other? When discussing AI and blockchain in domestic and international academia, the two are often separated, and few studies have put them together. The author believes that artificial intelligence and blockchain are the two sides of the intelligence revolution. The development of artificial intelligence will bring profound impact on the world's political and economic landscape, but at the same time, the development of artificial intelligence faces three major problems of privacy, security and fairness, and all three problems require solutions to be found in the blockchain. The basic line of this book is to start from issues such as AI risk governance and land the focus on the impact of blockchain on the future society. In the last part, the author also discusses the issue of intelligent civilization and intelligent social science. Intelligent civilization will become the new civilization model of human society after agricultural civilization, industrial civilization and information civilization, and on this basis, social science will also be newly reconstructed. For the first time in more than two hundred years, China stands on the threshold of the intelligent revolution at the same time as the developed Western countries, and this construction of intelligent social science will not only help the intelligent revolution to happen smoothly, but also become an important effort to construct an institutional discourse power in China.

I did not do this book on my own. Chapter 4, "Blockchain and Global Economic Governance Transformation" was completed by me and Zhang Jiteng. The other chapters of this book were all completed by me alone.

In addition, most of the chapters in this book have been published in national academic journals or well-known newspapers. The core of the introduction was published in *Liberation Daily*, November 26, 2019. In Chap. 1, the main content of

Q. Gao and J. Zhang, *Artificial Intelligence Governance and the Blockchain Revolution*, Artificial Intelligence and the Rule of Law, https://doi.org/10.1007/978-981-99-9211-9

Sects. 1 to 3 was published in *Contemporary World and Socialism*, No. 6, 2019; the main content of Sects. 4 and 5 was published in *Exploration and Free Views*, No. 1, 2019; and the core ideas of Sects. 6 and 7 were published in *Exploration and Free Views*, No. 10, 2017. In Chap. 2, the main ideas of Sect. 1 were published in *Liberation Daily* on August 1, 2019, the core content of Sect. 2 was published in *Chinese Social Sciences Today* on March 29, 2019, the main content of Sect. 3 was published in *Guangming Daily* on November 21, 2018, the main content of Sect. 4 was published in *People's Daily* on December 02, 2018. The core view of Chap. 3 was published in *World Economics and Politics*, No. 7, 2019. Chapter 4 is a collaborative effort between Zhang Jiteng and me, and the vast majority of its ideas were published in *Academics* 2019, No. 9 and *Global Reviews* 2019, No. 5. In Chap. 5, the core content of Sect. 1 was published in *Study Times*, August 23, 2019, the content of Sects. 2 and 3 was published in *Exploration and Free Views* No. 3, 2020. The main content of Chap. 6 was published in *Journal of Shanghai Jiaotong University (Philosophy and Social Sciences)*, No. 2, 2020. I would like to give special thanks to the editors-in-chief and editors who helped me to publish these results. They include Prof. Yuan Zhengqing, Associate Editor-in-Chief of *World Economics and Politics*; Prof. Peng Qinglong, Editor-in-Chief of *Journal of Shanghai Jiaotong University (Philosophy and Social Sciences)*; Editor-in-Chief Ye Zhudi and Deputy Editor Du Yunquan of *Exploration and Free Views;* Editor-in-Chief Hu Jian of *Journal of Social Sciences*; Director He Minjie and Editor Zhang Yao of the Theory Department of *People's Daily*; Editor-in-Chief Ma Lizhao and Editor Liu Liu of *Academics*; Deputy Director Wang Zhen of the Theory Review Department of *Liberation Daily*, Editor Ju Xiaojun of *Contemporary World and Socialism*, Deputy Editor-in-Chief Song Wenhao of *E-Government*, Editor Hu Xiurong of *Chinese Cadre Tribune*, Deputy Editor-in-Chief Sun Zhenhai of *Global Reviews*, and Editor Zhang Dandan of *Study Times*.

I would like to express my special thanks to my four mentors: Prof. Yu Keping, Prof. Shen Dingli, Prof. Li Luqu and Prof. Ding Jianshun. Mr. Yu is one of the most prestigious scholars in the field of political science in China, and he was my co-supervisor for my second postdoctor. Whenever he met me, Mr. Yu would encourage me to think more on political philosophy, which motivated me to go as deep as possible into the philosophical level when thinking about AI. Mr. Shen was my supervisor during my PhD. Mr. Shen's global vision and strong sense of mission would encourage me to think about the strategic impact of AI on the international landscape and global governance. Mr. Li was my advisor during my master's degree and the first postdoctoral period. Mr. Li's deep knowledge in the field of comparative politics encourages me to understand the new meaning of AI in different political and cultural contexts. Mr. Ding is my lifelong art teacher. Mr. Ding's excellent attainments in the field of Chinese humanities and arts, as well as his approachable mentoring style, have helped me understand more deeply that "politics is not only a technology, but also an art" when thinking about complex issues such as artificial intelligence and blockchain. I would also like to express my special thanks to Mr. Xu Dahua. Mr. Xu is a person very important in my life. Mr. Xu especially emphasized the comparison between Chinese civilization and Greco-Roman civilization, which made me realize more deeply the great historical significance of the development of AI to the change of human society and civilization.

The author would also like to express his gratitude to the leaders of East China University of Political Science and Law in all aspects. Secretary Guo Weilu, President Ye Qing, Vice-Secretary Ying Peili, Vice-Secretary Min Hui, Vice-Secretary Tang Bo, Vice-President Chen Jingying, Vice-President Zhang Mingjun, and Vice-President Zhou Lizhi of East China University of Political Science and Law have given me very much guidance and help in my work. The leaders of ECUPL are very concerned about the overall impact of AI on social sciences, and have established the Institute of Artificial Intelligence and Big Data Index at East China University of Political Science and Law, which has been assigned to me. This is both an affirmation of my previous AI research work and a new heavy responsibility. The leaders of ECUPL's functional departments and schools, such as Director Qu Yuliang, Director Dai Ying, Director Liu Danhua, Director Xia Fei, Director Yang Zhongxiao, Dean Hong Dongying, Director Zhou Litiao, Director Han Qiang, Director Qu Wensheng, Director Sun Liming, Director Hu Ye, Dean Chen Jinzhao, Director Cui Yongdong, and Secretary Que Tianshu, have helped me a lot. I would like to express my heartfelt thanks to all of them.

I would like to thank my colleagues in the computer science program at ECUPL for their support and assistance in my AI research. I would like to especially thank Prof. Wang Yongquan Wang, Prof. Wang Yi, and Mrs. Liu Yang. With the full support of several teachers and fellow computer science majors, we began to explore the technical practice of counseling robots, which allowed me to personally practice the possibility of using AI technology to change society in addition to my research in the social sciences. We are grateful to the team of the Institute of Political Science of ECUPL, including Associate Professor Wang Jinliang, Associate Professor You Tengfei, Associate Professor Yan Xingjian, Lecturer Ji Lei, Lecturer Zhu Jian, and Lecturer Du Huan. Our institute is like a young family. Supported by ideals and beliefs, and under the atmosphere of unity and tension, we all move forward happily in spite of difficulties. Our doctoral and master's students are also members of this big family, and they undertake a lot of administrative work and data compilation in the institute. These graduate students mainly include Zhou Rongchao, Jin Hua, Cai Congyu, Que Tiannan, Lv Junyan, Zhao Qiao, Yang Fan, Zhang Jiteng, Yang Yuxiao, Shuyu, Meng Nuoyi, Tang Mengnan, Mo Fei, Jia Yilin, Liang Mingyue, Sun Jiemei, Yang Jiao, Wang Jinrui, and Liu Zhiyou. They saved me a lot of time in finding and proofreading the literature of the book. Special thanks go to Shu yu and Mo Fei, who drew the mind maps and structure diagrams for the chapters of the book.

I would like to give special thanks to my good friend Prof. Gu Yu. Mr. Gu is an expert in the field of the history of Chinese political thought. Mr. Gu has inspired me in my studies mainly in the field of Chinese studies. While the knowledge of Chinese studies is closer to the Qi Sect, AI and blockchain are closer to the Sword Sect, and I think using Qi to protect the Sword is the correct approach.

Thanks to my wife, daughter and parents. My wife, Zhang Xianli, has recently developed an extremely strong interest in philosophy, and often quotes Democritus,

Epicurus, Wittgenstein, Kant, Hegel, and so on. To be honest, I am extremely enjoyable. I want to be immersed in the sea of knowledge every day. Interaction with my wife is also a constant learning experience, which is indeed the enjoyment of life. My wife is not only very devoted to academic research, but also helps me with a great deal of family affairs, which allows me to soar very freely in the enjoyment of the joy of knowledge. My daughter, Gao Mohan, is now eleven years old, and I feel that she is growing up every day. I used to see her perspective as that of a child or a teenager, but recently I realized that it is not. Because her generation lives more under electronics, they have a more natural understanding of artificial intelligence and big data. Or perhaps her generation understands true AI better than we do. I am grateful to my parents who gave birth to me and raised me, and who have always supported me silently. My father has very insightful views on many macro issues that have deeply influenced me. My mother is particularly motivated, and after her retirement she taught herself cucurbit flute and taijiquan, practicing hard and diligently every day, a spirit that inspires me all the time.

In the process of editing and publishing this book, I received great help from Deputy Editor-in-Chief Cao Peilei of Shanghai People's Publishing House, and Editior Feng Jing of Shanghai People's Publishing House. I have benefited greatly from their macroscopic vision of academic issues, their rigorous editing attitude, and their demand for precision in writing.

During the English translation process of this book, colleagues such as Qiao Lijuan, Liang Zihan, Sui Xiaozhou, Tai Runze, and Zhang Haomiao helped me proofread the manuscript. The large number of charts in the manuscript also consumed a lot of energy from everyone.

I would like to express my heartfelt thanks to all the teachers, leaders, colleagues, friends and family members who have given me support and help!

Gao Qiqi
March 1 2020